Taxonomy of Bamboos

Taxonomy of Bamboos

P. E. Bedell

2024

Associated Publishing Company®

A Division of

Astral International Pvt. Ltd.

New Delhi – 110 002

First Published, 1997

Reprinted, 2024

ISBN: 978-81-86580-29-5 (HB)

Published by : **Associated Publishing Company®**
A Division of
Astral International Pvt. Ltd.
– ISO 9001:2015 Certified Company –
4736/23, Ansari Road, Darya Ganj
New Delhi-110 002
Ph. 011-43549197, 23278134
E-mail: info@astralint.com
Website: www.astralint.com

Printed at : Replika Press Pvt. Ltd.

FOREWORD

Bamboos are some of the most important forest products used by people in Asia and the Pacific. They have amazingly high growth rates, multiple uses and constitute the most important long fibred raw material for the Indian pulp and paper industry. The phenomenon of their gregarious flowering and consequent enmass drying are the biggest problems faced by Indian forest management and paper industry. Bamboos are important both for economic as also ecological considerations.

Earlier attempts to classify bamboos by the conventional system based on flower and fruit characteristics resulted in some bamboo species with wide geographical distribution and termed as separate species, though they were polypoids of the same species. The resulting confusion in nomenclature and difficulties in identification indicated that only a multidisciplinary approach would help to bring about correct identity and rational bamboo classification. Thus, the author's attempt at presenting the body of evidence from other related disciplines for a natural classification of bamboo is highly commendable.

She has described in general terms the nature of plant taxonomy citing appropriate examples of bamboo species. The book virtually presents Taxonomy as a contemporary science using modern methodology and principles of classification and will prove helpful in paving a way to ultimate solution of the long standing problem of bamboo classification.

I am confident that this book on 'Taxonomy of Bamboos' will prove useful to researchers, professional foresters, university students of Botany and also to bamboo cultivators. I congratulate Ms P.E.Bedell for bringing out this book.

R.B. Kale (I.F.S.Retd.)
Forestry Consultant
68, Nilgiri Apartment, Alaknanda 'A'
New Delhi 110019

PREFACE

Bamboos are unique in more ways than one. For instance they grow to their full height in one season unlike other trees. Further, although they belong to a predominantly herbaceous family-Gramineae, they differ from other members in being mostly woody, and yet is different from other woody trees. Then again, unlike other members of the plant kingdom, their flowering and seeding rhythm is unpredictable, and many of them flower at long intervals about once in a life time in 30 to 60 years. But with all these unorthodox characteristics, they are yet one of the most valuable group of plants, and has fascinated man from the beginning. And yet they have remained an enigma to Taxonomists who are responsible for classification and nomenclature of all plants.

Bamboo research is therefore a field of Scientific investigation which has attracted students and researchers from diverse academic disciplines. New research work on the subject has been published in a perplexing diversity of publications and is inevitably concerned with the minutiae of the subject. The sheer number of these publications also causes confusion and difficulties of assimilation. Further, there is as yet no substantive literature on Bamboo Taxonomy and no clear conceptual framework within which students may function. Under these conditions students are usually cast adrift in a vast amount of literature on miscellaneous aspects of Bamboo structure, utility, distribution, and growth characteristics. There is thus a need for a comprehensive presentation of those areas of research which are dealt with piecemeal in research papers. This will enable College and University students in Botany, and students in Forestry, to go on from this text, to read scholarly research articles with profit. This book is designed to satisfy this need. I also hope that this book will enable Foresters and Scientists to glimpse possible areas where they may cooperate with mutual profit. To this end the research findings in various disciplines have been introduced and integrated.

The book is the result of thirty five years of deep interest and dedicated research on various aspects of the subject, during which time many professionals, friends, and acquaintances have given me a helping hand which is acknowledged herewith. Il am grateful to the Government of India for the scholarship I received for this study from June 1961 to September 1965, and to the Forest Research Institute, Dehradun for providing me with all the facilities required. I am also grateful to Mr. K. Ramesh Rao, the then Branch Officer of the Wood Anatomy Branch, FRI, Dehradun for his help and guidance, and to Mr. T.N. Srivastava, IFS, the then President of the FRI & Colleges, and to Mr. I.M.Qureshi, IFS, the then Director of Forestry Research, FRI, Dehradun for taking an active interest in my work on Bamboos. I am also grateful to Ms. Nazma Sultana, the then Publicity Assistant and now the Central Librarian, FRI, Dehradun for helping me keep abreast of literature and to the various experts on bamboos I have met and corresponded with from time to time, especially Prof. Walter Liese of Germany who happened to visit FRI, Dehradun during the time of my scholarship and for his useful suggestions. Further, I am grateful for the various help and support I have received in gathering and classifying the information for this book from my friends Ms Esther Chawner and Ms Margaret Houison, and most of all to my father Mr Pattanath Gopalan, and my husband Maj Desmond Edwin Bedell, both of whom have always been a source of great inspiration to me and have helped me analyse and see and put things in their correct perspective.

P.E.B.

CONTENTS

LIST OF PLATES

LIST OF TABLES

1

TAXONOMY

GENERAL CONSIDERATIONS

Taxonomy is the theory and practice of classifications and classification of plants is important for recognition, delineation, and orderly arrangement of distinct entities encountered as ecological units, as constituents of forest floras, and as potential sources of raw materials. Most taxonomists strive to produce 'natural classifications', that is, classifications based on consideration of as many features of the organisms as possible with a view to reflect their evolutionary relationships.

All plants have a common cytogenetical foundation stemming in part from the ubiquitous genetic material—the DNA. But different species have had different evolutionary opportunities, and in many instances they have utilized their common genetical and chromosomal endowment to adopt different evolutionary strategies which are reflected in their patterns of variation. Thus the relationship which is traced to a common ancestor is described as evolutionary or phylogenetic relationship.

There is also a relationship in terms of some shared features, regardless of the reason for the sharing based simply on the resemblances which they show in the features studied. Such relationships based on perceived similarities are described as phenetic. For example chemical analysis may show that a number of different species are all very similar in terms of polysaccharides stored in their seeds. This would then reveal a phenetic relationship, and not necessarily a phylogenetic one. The distinction between phenetic, phylogenetic and natural classifications are discussed in detail by Davis and Heywood (1963), Heywood and McNeil (1964) and Heywood (1967).

As opposed to 'Natural Classifications' is the 'Artificial Classifications' based only on one or at the best a few characters. This may be purely a phenetic relationship and usually have a limited application in keys for identification. Bamboo research on classifications so far have been directed towards finding similarities and differences solely for the purpose of distinguishing different bamboos from each other. But the complexity of structure and behaviour in bamboos could provide much evidence on which to base a natural classification, if a proper systematic and detail study is made. The groups recognized in this sort of classification are likely to arise from a common ancestor.

Taxonomists have always felt more confident about characters which can be associated with a function. As such they depended in the beginning purely on comparative morphology of flowers, inflorescences, leaves and fruits. Later, comparative anatomy was also adopted. The authors of Die Naturlichen Pflanzenfamilien were perhaps the first to apply anatomical characteristics to overall classifications on a grand scale.

The contribution of Jeffrey, Bailey, Eames, and Sinnot, to systematic anatomy are well known. They laid the pattern for an overall classification of vascular plants on morphological anatomical grounds. The findings of Irwin Bailey and his Associates with regard to the development, form and sequence, of vessel types in vascular plants have shed a clear cold light on the relationships of gymnosperms with angiosperms and of monocotyledons with dicotyledons.

In the work of Bailey and his Associates (1954) on the Ranalian Complex is found an excellent model showing what can be done to clarify structures and relationships. In many instances a re-evaluation of previously worked families and genera has yielded extremely valuable information. One point which stands out above all others in their findings, is that it is important to assess the data obtained from all parts of the plant in a combined way. The vulnerability of conclusions drawn from over emphasis of the data from a single structure of a group of plants has been clearly demonstrated, and no one should fall into such a trap today. (Collins 1957)

At the turn of this century, taxonomists have turned more and more to other disciplines for obtaining data on evolutionary relationships of enigmatic taxa. It is now well known, that where classical systems of taxonomy are less well founded as in bamboos, nucleic acid comparisons for instance may provide a source of taxonomic evidence. The DNA is a nucleotide polymer which is the physical manifestation of the genetic code, and all individuals have a genetic code unique to themselves. Furthermore, though much remains to be discovered about the genetic code, it is now a recognized fact that nucleic acids of different organisms are treasures of taxonomic evidence, because DNA is the language in which specifications for the form and physiology of an organism is written, and the genetic code is the statement of its evolutionary history minimally modified by the environment. Modern methods now enable us to unravel this code. Similarities between genetic sequences of individuals betray community of descent; and translation of the protein structure, will appeal to all systematists as a source of Phylogenetic information, in addition to acting as taxonomic evidence.

The properties of proteins are diverse, enabling the use of different analytical techniques for their investigation. Amino acids are amphoteric, containing a basic amino group and an acidic carboxyl group. Many contain more than one such groups not involved in the peptide linkage which are hence free to ionize in different pH conditions. Proteins will therefore have a net positive or net negative charge which will vary with the pH of the medium. These electrical properties can be used to separate different proteins which have different net charges by the technique known as electrophoresis. Variations of this method are very often used in protein separations for taxonomic purposes.

Sequence analysis of proteins rests upon the ability of certain reagents and enzymes to break peptide bonds selectively, that is only those between particular amino acid residues. The enzyme trypsin for example selectively cleaves the lysine-arginine bond thus providing tryptic peptides. These peptides may then be separated by chromatography and electrophoresis in the technique known as 'finger printing'.

The pattern of peptides (or 'finger printing') can be used as a source of taxonomic evidence. Similar proteins indicating individuals of a species are assumed to have lysine-arginine bonds in similar positions and so to produce peptide patterns which are similar. If peptides are separated in a pure form (usually by chromatographic procedure) they can be selectively hydrolysed from either the free amino group end, or the free carboxyl group end, by various reagents. As each terminal amino acid is hydrolysed free, it can be identified by chromatography, and so the sequence of the different peptides can be established.

Further, protein molecules vary considerably in size and this can be used as a criterion for their fractionations. The pattern of fractionation may be potentially useful in taxonomy. Ultracentrifugation of proteins is a separation technique directly related to the size of their molecules. This method has been used to study proteins of the Leguminoseae and Graminae by Danielsson (1949). He found two proteins which he termed as legumin and vicillin which were recognisably distinct in terms of size in most of the genera listed and the proportion of the two types varied from genus to genus.

There is also a growing recognition of serotaxonomy as an aid to phylogenetic investigations. This happened with the discovery that all antibodies and many antigens were proteins, often specific to the organism producing them. Kowarski, Bartacelli, Magnus, and Rives, are some of the early notable plant serologists who have made significant contributions to the application of serology to systematics. Further, Nelson and Birkeland (1928) compared wheat strains of known genetic relationship by means of serology and found that serological data correlated well with the known phylogeny of wheat varieties.

Thus the widespread use of chemical data as taxonomic characters, marks a recent extension of the range of recognized sources of taxonomic evidences such as morphology, anatomy, embryology, cytology, ecology, and genetics.

GENERAL CHARACTERS OF BAMBOOS

Bamboos are woody monocots and belong to the family Gramineae. They differ greatly in stature and form and vary from lofty forms attaining a height of 30 metres or more in species of *Bambusa* and *Dendrocalamus* to mere undershrubs as in species of *Arundinaria* found mostly at high altitudes. Although most species have erect woody culms, there are some like *Teinostachyum helferii* that are straggling or semi scandent, and others like *Dinochloa maclelandii* which are truly scandent, and climb into the crowns of nearby trees. They generally occur in clumps which may be tufted as in *Bambusa arundinacea* or single stemmed as in *Melocanna bambusoides*. Culms arise from an underground root stock known as the rhizomes consisting of a twisted mass of entangled branches producing a large number of closely packed culms in the case of tufted species. These densely packed culms forming clumps are described as sympodial. On the other hand in single stemmed species, rhizomes have long creeping branches, and culms are produced 30 to 90 cm apart, and are described as monopodial. Rhizomes bear buds and roots. Buds are initially flat in shape, and are covered profusely with overlapping scales. It is the rhizome-buds which eventually develop into culms and the overlapping scales into sheaths, when they are a year or two old.

The culms may be hollow throughout their lengths, interrupted at intervals by nodal plates as in *Bambusa nutans, Dendrocalamus longispathus* etc., or they may be partly solid and partly hollow as in *Bambusa tulda*. However, whether solid or hollow, they are divided into prominent ridged nodes, and smooth long cylindrical internodes bounded on the outside by a persistent epidermis. The internodes increase in length from bottom upwards till about quarter or mid height, and then decrease in length again towards the top (pl 1). The culms in all cases taper from the bottom upwards, and vary in diameter and wall thickness from species to species. They also vary in the number of internodes and the maximum and minimum length of internodes, in different species, and to a certain extent within individuals of a species. Furthermore, in some species, the lower nodes have a ring of dormant spine like rootlets of 1 to 2 cm in length. The nodal plates may be straight, curved, convex, or concave, depending on the species. Branches are borne only at the nodes. In some species as in *Bambusa arundinacea*, branching starts from the lower nodes, and in others such as *Thyrsostachys oliveri*, branching starts from about quarter to mid height. Branches may appear singly or in groups of three to five at each node on alternate sides of internodes. When arranged in groups of three or five or more, there may be a main branch thicker than others, or all the branches may be of the same size. The branching pattern varies with the species.

Leaves are distichous and consist of a tubular sheath split at the base. Leaf blades are generally flat and may be linear-oblong, or lanceolate, with a midrib and numerous longitudinal veins which are generally of two kinds—the stout and the slender. Between two stout veins are five to nine slender veins. All bamboo leaves have transverse veins which connect one longitudinal vein with the other. These transverse veins vary from straight to oblique to a bend in the middle in different species. The identification of many species according to Gamble (1881) is facilitated by counting the number of longitudinal nerves in an unit area. In almost all bamboos, the leaf blade is joined to the leaf sheath by a short petiole.

The younger culms are covered by culm sheaths which consist of a lower portion known as sheath, and upper portion known as the blade. The relative sizes of the sheath and blade varies in different species. Between the sheath and the blade are the ligules, and the auricles. The blades, ligules, and auricles vary in shape, structure, size and orientation in different species. The culm sheaths are borne at the nodes, and may cover the entire internode, or may extend only half way up the internode, depending on the species. The sheaths also vary in colour and texture in different species, and may be striated or plain, or hairy, depending on the species.

Flowering in bamboos is irregular and unpredictable in most species, and they may flower only once in a life time. The flowers are much reduced in size, and consist of scale like structures known as lemmas and paleas, and are arranged in spikelets which vary from species to species. The spikelets may consist of one to many flowers arranged in panicles, racemes, dense clusters or terminal heads on a short or minute slender axis known as the rachilla. Flowers are usually bisexual, sometimes unisexual, small and inconspicious, consisting of stamens and pistil, and 2 or 3 hyaline or fleshy scales known as lodicules representing the perianth, subsessile between

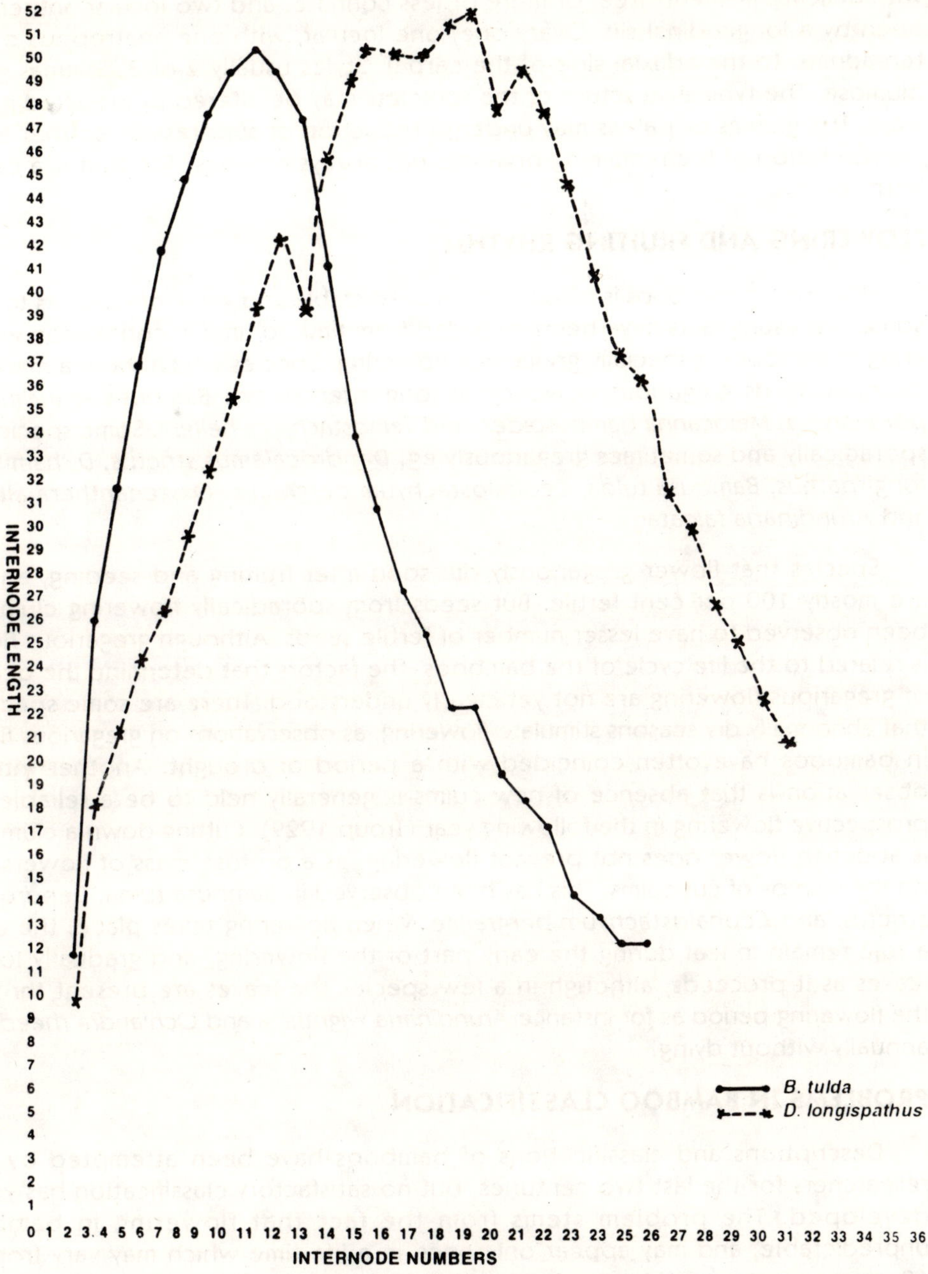

Pl. 1. Variation in internode length within the culm in *Bambusa tulda* and *Dendrocalamus longispathus*

two bracts—the lemma and palea, the whole forming a floret or false flower. The paleas are 2-keeled or keelless or suppressed. The florets bear at the base two empty bracts (the upper and lower glume), Stamens are hypogynous 3 to 6 rarely more, with delicate filaments free, or more or less connate, and two locular anthers, which open by a longitudinal slit. Ovary one, one locular, with one anatropous ovule, often adnate to the adaxial side of the carpel. Styles usually 2 or 3. Stigmas generally plumose. The typical structure of the spikelets may be altered or modified in various ways. The glumes or paleas may undergo reduction or suppression so that a correct interpretation of the remaining organs is not always obvious. The fruit is a caryopsis, berry, or nut.

FLOWERING AND FRUITING RHYTHM

Flowering in bamboos is unique and unpredictable and vary from species to species. Some interesting facts have been recorded from time to time in connection with flowering in bamboos, especially gregarious flowering. Species which show a marked tendency towards gregarious flowering at long intervals are *Bambusa arundinacea, B. polymorpha, Melocanna bambusoides,* and *Teinostachyum helferii.* Some species flower sporadically and sometimes gregariously eg. *Dendrocalamus strictus, D. hamiltonii, D. longispathus, Bambusa tulda, Cephalostachyum pergracile, Oxytenanthera albociliata* and *Arundinaria falcata.*

Species that flower gregariously die soon after fruiting and seeding, and seeds are mostly 100 per cent fertile. But seeds from sporadically flowering clumps have been observed to have lesser number of fertile seeds. Although gregarious flowering is related to the life cycle of the bamboos, the factors that determine the occurence of gregarious flowering are not yet clearly understood. There are some speculations that abnormally dry seasons stimulate flowering, as observations on gregarious flowering in bamboos have often coincided with a period of drought. Another interesting observation is that absence of new culms is generally held to be a reliable sign of prospective flowering in the following year (Troup 1929). Cutting down a clump which is about to flower does not prevent flowering, as a profuse mass of flowers appear on the stumps of cut culms. This has been observed in *Bambusa tulda, Dendrocalamus strictus,* and *Cephalostachyum pergracile.* When flowering takes place, the culms as a rule remain in leaf during the early part of the flowering, and gradually lose their leaves as it proceeds, although in a few species the leaves are present throughout the flowering period as for instance *Arundinaria wightiana* and *Ochlandra rheedii* flower annually without dying.

PROBLEMS IN BAMBOO CLASSIFICATION

Descriptions and classifications of bamboos have been attempted by various researchers for the last two centuries, but no satisfactory classification has yet been developed. The problem stems from the fact that flowering in bamboos is unpredictable, and may appear only once in a life time which may vary from 30 to 60 years or more. As the orthodox system of classification is based on the reproductive structures, the early botanists had perforce to depend on herbarium specimens for taxonomical studies of bamboos.

Furthermore, the flower structure in bamboos as in all members of the Gramineae are so reduced, it needs specialized methods of observation such as the use of low power microscope for fresh specimens, and the use of serial microtome sections for observations of herbarium specimens. These two factors viz nonavailability of flowers as and when required, and the reduced size of the flowers, have led to considerable confusion regarding classification and nomenclature of bamboos.

The later Botanists such as Kurz, who studied living specimens in the field, observed vegetative characters which differed considerably from species to species, and recommended them for developing a system of identification and classification. Thus researchers like Gamble (1881) and Brandis (1896-1921) made an intense effort to study living specimens in detail. Later in this century, researchers in different countries took up studies on different aspects such as morphology of the culm sheaths (Raizada and Chatterjee 1963), morphology and anatomy of the culms (Pattanath 1965) anatomy of the culms (Shigenmatsu 1985; Ghosh and Negi 1960; Pattanath and Rao 1969; Grosser and Liese 1971; 1973; Grosser and Zamuco 1971; Kitamura et al 1974; Wen Taihui and Chouwenwai 1985), leaf characteristics (Metcalfe 1956, Fujimoto 1966), culm buds and bud sheath (Bahadur 1979); cytological studies (Richharia and Kotwal 1940; Parthasarathy 1946; Janaki Ammal 1959; Zhang Guangzhu 1985). The approach individually was piecemeal rather than the whole plant, and hence remained phenetic and solely for the purpose of identification, rather than classification.

Moreover, because descriptions of bamboos so far have been based on a very limited number of specimens which are insufficient to give a full picture of their variation, population structure, and distribution, many species have been placed under the wrong group, for instance a Burmese bamboo known as tabindiang in Burmese, and wagbai in Karen, was observed by Brandis to resemble *Gigantochloa, Oxytenanthera* and *Melocanna* and yet differed from all three of them. In another instance floral specimens of *Gigantochloa macrostachya* collected by Brandis in 1862 on further examination convinced him that it was *Oxytenanthera* and should be called *Oxytenanthera macrostachya.* Dransfield (1980) believes that some bamboo species in Philippines which are described under *Schizostachyum* should belong to a different genus. In still another instance Holttum (1956) suggested that *Cephalostachyum* and *Teinostachyum* should be included in the genus *Schizostachyum.* Thus the problem of bamboo classification is even now far from a final solution.

This situation has arisen mainly due to the difficulty of delimiting bamboo species, and determining the discontinuities in variations. It thus calls for a consideration of what constitutes a species. Firstly, species are distinct from one another in recognizable ways. Secondly, variation between species is discontinuous, that is, they are not connected with one another by intermediate forms, although within a species, variations between individuals do occur. This variation may not be uniformly distributed, but may give rise to local intraspecific populations that are recognizably distinct and which may be termed as races. Nevertheless, the component races intergrade, and within the species, variation is more or less continuous, and there are no sharp breaks. Finally individuals of a species are interfertile with one another. Therefore a species can be defined precisely as a series of similar intergrading and interfertile populations recognizably distinct from other such series, by genetically controlled barriers, preventing interbreeding. The limits of bamboo species have not so far been experimentally studied, but are merely implied from apparent discontinuities in variations. Therefore the limits have been wrongly drawn, either too widely or too narrowly. Moreover, the available evidence so far, is very incomplete and quite inadequate. Thus different scientific names have been applied to what later observations have shown to be completely different genera, or two different genera have been mistaken to belong to one genus.

For instance *Oxytenanthera monostigma* Bedd was known as *Bambusa ritchey* Munro. In another instance McClure (1973) transferred the genus *Guadua* of tropical America to the genus *Bambusa*. There are about 30 recognized species of *Bambusa* in tropical Asia which are diverse in vegetative characteristics and flowering behaviour. But all of the them have one feature in common and that is, many flowered spikelets. Two groups of *Bambusa* are recognized by some researchers, one having thin walled long internodes occurring usually at higher altitudes. They flower frequently and the clumps do not die after flowering and fruiting. The other group has species with thicker walls and shorter internodes and occur at lower altitudes. They flower infrequently and the whole clump dies after flowering. Holttum (1954) suggested that the first group should be transferred to another genus.

Some of these difficulties have arisen from the nature of evolution itself. The two components of species definition viz recognizable distinctness and reproductive isolation sometimes get out of setp. Thus two series of populations may be isolated from each other by a genetically controlled barrier so that they do not interbreed, but in most of the other characteristics they may appear to be indistinguishable. For example *Arundinaria khasiana* Munro and *Arundinaria falcata* Nees resemble each other closely. Only, in the former, culms are harder, leaves are broader and transverse veins are faintly visible. Then again *Arundinaria patlingii* Gamble and *Arundinaria griffithiana* Munro are very similar, only the former is doubtfully spinescent.

Sometimes differences between a series of species become obscured by the development of hybrids with two, four, six, or even eight times the number of chromosomes of the original parent species. Unless such a situation is appropriately investigated, it may be interpreted as the existence of a single, widely variable species. The existence of such series of related species with different multiples of the same basic chromosome number is known as polyploidy. The cytology of most of the bamboos is not yet well known. But many of the bamboos studied in the Forest Research Institute Dehradun, are tetraploids, and the genera *Dendrocalamus* and the Asian species of *Bambusa* are hexaploids (Varmah and Bahadur 1980). Somatic chromosome numbers for some of the species have been reported by Richharia and Kotwal (1940), Parthasarathy (1946), Janaki Ammal (1959) and Zhang Guang-Zhu (1985). Hybridisation combined with polyploidy is a common cause of taxonomic confusion especially when it has not been investigated experimentally (Jeffrey 1982). An example of the taxonomic problem caused by polyploidy has been illustrated by Jeffrey 1982 as follows:

APPARENT SPECIES

<------------------------- Range of Variation ------------------------->

At first sight we have apparently one species with a wide range of variation. On further investigation it turns out to be made up of a series of distinct diploid species (A.B. and C) of narrow variation range overlaid by the development of tetraploid AB and BC and hexaploid ABC, as shown diagramatically, below:-,

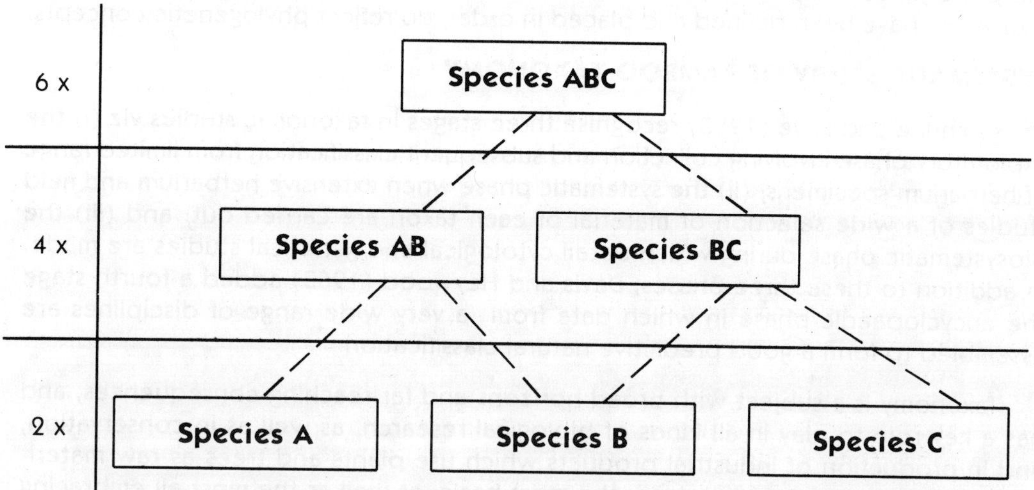

Yet another complication that can arise, stems from hybrids reproducing by asexual means, known as apomixis. Bamboo species may well be apomicts and they reproduce by rhizomes. The result of this kind of reproduction is often, the production of large numbers of individuals, which at first sight look very much like the individuals of a species, but differ from them in not interbreeding amongst themselves, and in not varying genetically from individual to individual. They represent as it were, cuttings of one original plant and all share more or less the genetic make-up of the ancestral parents. (Jeffrey 1982).

A common finding in research on bamboo classifications so far is that the generic characters are not too well defined and the generic classification is in doubt, and in a few cases a species has been placed in two or more genera by different researchers. Problems of bamboo classification arise fundamentally from the nature of variation patterns exhibited by the different categories of taxa. Numerous patterns of variations have been observed in different species of bamboos by different Researchers, and it is likely that many more remain to be discovered. Normally variation patterns are such as to enable grouping into a heirachial classification, but when they elude the heirachial system the problems arise from difficulties of interpretation of complex patterns of variation, originating from operation of curious breeding systems, such as apomictic, or self pollinating groups, or polyploid complexus, or populations of species which are hybridizing as already discussed.

Bamboos also appear to present special difficulties in representing variation in heirachial terms. This is mainly due to difficulties in circumscription of groups within the variation spectrum, that is, the requirement for lines to be drawn around 'nodes' of variation to represent species, genus etc. The problem here is to know where exactly the lines should be drawn. Though most taxonomists classify plants in accordance with the maximum correlation of attributes, and do not in principle consider the phylogenetic relationships of the materials they work with, yet eventually, the taxa which have been defined and placed in order, do reflect phylogenetic concepts.

SYSTEMATIC STUDY OF BAMBOO TAXONOMY

Valentine and Love (1958) recognise three stages in taxonomic studies viz (i) the exploratory phase involving collection and subsequent classification from limited range of herbarium specimens; (ii) the systematic phase when extensive herbarium and field studies of a wide selection of material of each taxon are carried out; and (iii) the biosystematic phase during which detail cytological and genetical studies are made. In addition to these three phases, Davis and Heywood (1963) added a fourth stage the encyclopaedic phase in which data from a very wide range of disciplines are assembled to form a good predictive natural classification.

Taxonomy is a subject with broad horizons and far reaching consequences, and has a key role to play in all kinds of biological research, as well as in conservation, and in production of industrial products which use plants and trees as raw materials. It is at one and the same time, the most basic, as well as the most all embracing of all biological sciences. The concept of taxonomy has changed much with the passing years, with changing technology, and with comming into prominence of new fields of investigations such as genetics, molecular biology, and electron microscopy.

In taxonomical studies of bamboos, the first two phases viz the exploratory phase, and the systematic phase have received the most attention so far. It is only recently that the third phase viz the biosystematic phase has received some attention. The work on the fourth phase viz the encyclopaedic phase has hardly received any attention at all.

The earliest attempts at the exploratory phase, dates way back to 1750 when Rumphinus described some bamboos, followed by the work of Schreber (1789), Blanco (1837), Ruprecht (1839), and Munro (1868). The second phase viz the systematic phase began with Kurz in 1876 when he suggested that vegetative characters of bamboos should be studied from living specimens as they would be useful in the classification of bamboos. His suggestion was taken up by Gamble (1881), Bentham (1883), Stapf (1897), Hooker (1897), Camus (1913), Brandis (1921), Agnes Arber (1926) and Holttum (1946-1956) who made extensive and detail studies. These studies were soon followed up by a number of researchers from all over the world working on different aspects of bamboos. But all these works were carried out as isolated, independent items of research and there was no coordination. It was only in late nineteen seventies and eighties that coordinated work on the systematic phase began, and germplasm collections of bamboos were initiated. There are at present a number of germplasm collections of bamboos in China, India, Nepal, Thailand, Sri Lanka, and Africa mainly as living collections in botanical gardens and arboretums.

As already mentioned, the work carried out in the biosystematic phase is relatively recent and restricted to studies on chromosome numbers, and a few hybridisation studies. The bulk of information available on this phase, is on Chinese bamboos. Zhang Guang-Zhu (1985) reported chromosome numbers of sixteen species of Chinese bamboos and made an attempt to correlate chromosome numbers of bamboos with their adaptation to variations of temperature zones. Further, Zhang Guang-Zhu and Chen Fu qui (1985) reported some interesting results on bamboo hybridisation studies, which confirm the presumption that problems in bamboo taxonomy result from their curious breeding systems.

All these factors emphasise the need for a more detail study with a larger number of specimens covering the entire range of the variation spectrum of each species. Further, as bamboos seem to elude any one specialized line of analysis, it is clear that they require a wider horizon, and a multi-disciplinary approach for them to yield to being characterized, grouped, and classified. This will make the classification as natural as possible, and minimise errors due to possible convergent evolution, as the number of attributes chosen for comparison would then be multiplied. The presumption is that no two taxa will be convergent in all their features. Thus, though vegetative characters and floral morphology may appear to be similar, there may be marked differences in anatomical and cytogenetic characteristics. The enormous number of attributes and attribute states, that this multidisciplinary approach requires would eventually need a computer analysis of similarities and differences.

2

CLASSIFICATION

GENERAL CONSIDERATIONS

The basic process in classification is identification followed by the naming of the specimen plant. Identification is the process of deciding whether two individuals resemble each other sufficiently to be regarded as the same. This is the basic principle of all classifications. Though in theory this may seem simple, it is accompanied by a number of difficulties in practice, because of the nature of variation in the plant kingdom caused by evolution.

For instance most of the bamboo species are known only from a very few specimens which are insufficient to give a full picture of their variation, population structure, and distribution. As a result, many species still remain undescribed and unnamed. And some have been described more than once and considered to be a different species especially when specimens have been obtained from the extremes of the range. Discontinuity of variation which is caused by divergence and extinction during the course of evolution is the main point which enables us to sort plants into distinct groups. The framework that has been built up for this purpose in scientific classification of plants is known as the taxonomic heirarchy, and is made up of a set of conventional taxonomic categories arranged in a conventional order, which must be strictly adhered to. The level at which a category stands in the heirarchy is known as its rank. The groups of plants themselves are known as 'taxa' in the plural and 'taxon' in the singular.

Taxa of lower rank are always subordinate to and included in those that are of higher rank. For instance a species is made up of a number of plants. A number of species which have some similar characteristics are grouped together under the category genus, which is a unit of higher rank than the species. Therefore a genus as a whole, would show a much greater range of variation than does any one species within the genus. The limit set to a taxon can be expressed either descriptively or in terms of its range of variation.

The species therefore are grouped together into genera, and genera into families, and families into orders, and so on. Below the species is a series of subordinate categories such as subspecies, varieties, form and so on. In the classification of bamboos there are also two categories between the genera and the family termed as tribe and subtribe. At the lower levels of taxonomic heirarchy, classification becomes more detailed and specific, about the information on the plants concerned. Conversely at higher levels such as tribes, family, order etc. classification becomes more generalized.

Species is considered as the basic unit of classification because they are the only taxa for which limits can be defined and experimentally verified. This is so in sexually reproducing outbreeding plants. Above the rank of species what constitutes a taxon, is solely a matter of opinion. There is no accepted biological or taxonomic definition of what should be considered as a subtribe, tribe, or family, and there is no objective method of proving anyone's opinion as right or wrong. But all the same, it is a matter of considered opinion of experienced and competent taxonomists, and is based on certain principles of delimitation. Thus all members of any taxon should resemble one another in the sum total of their characters more closely than they resemble the members of any other taxon of the same rank. For instance all species included in the genus *Bambusa* should resemble each other more closely than with the species included in the genus *Dendrocalamus.* Then there should be a clear break, that is, a comparative discontinuity in variation. Further, the discontinuity between two genera should be greater than the discontinuity between two species. The delimitation and arrangement of taxa above the rank of species is an attempt to construct a more natural classification at different levels of taxonomic heirarchy. In this lies the disparity between the artificial classification (which takes into account only a few features such as gross visible appearance) and a natural classification which takes into account all biological evidence.

The placing of an imperfectly known material may be done more easily in an artificial classification than in a natural classification. But a poorly known plant that has been placed in one position in artificial classification may have to be shifted to another position when more of its characteristics get revealed. A shift of this kind is known as a change in taxonomic position, and such changes become necessary when further information on the species reveals that the plant has been hitherto misplaced in classification.

Identification of the plant specimen is followed by the process of naming it. The part of systematic botany that deals with the giving of names to plants is known as nomenclature, and this is a part distinct from identification and classification. It is only when a plant has been identified to belong to a taxon, that it can be given a name. Scientific names must avoid all confusion and serve the purpose of easy means of reference, unlike common names which may mean different things to different people when referring to one and the same thing.

In order to fulfill this condition, the rules of botanical nomenclature has to be followed which is known as the International Code of Botanical Nomenclature (ICBN). It governs the formation and usage of all scientific plant names except those of cultivars. A fundamental provision of this code is that scientific plant names must be in Latin. This is to avoid multiplicity of names in different languages for one and the same plant. Latin has been chosen as the language for naming plants because of its early usage in the science of botany, and also because it is a dead language which is no longer used as the native tongue by any one group of people.

Plants are usually named at the species level, and the name of the species is made up of two parts viz the name of the genus to which it belongs followed by the second part known as the specific epithet. For example *Oxytenanthera nigrociliata* is the scientific name of a bamboo which belongs to the genus *Oxytenanthera* and *nigrociliata* is the specific epithet. The system of naming species by means of a generic name plus a specific epithet was established by the Swedish Botanist Linnaeus (1707 - 1778) and is known as the binomical system. The generic name is always written with a capital initial letter. The scientific name of a plant is followed by the abbreviated form of the name of a person eg. *Oxytenanthera monostigma* Bedd. This is the authority citation and the person is the author of the name of the plant concerned, i.e. the person who first published the name. His name is not part of the botanical name but is added for the purpose of precision. The problems in botanical nomenclature are (a) how to reconcile the need to accommodate changes in the taxonomic system, (b) the need for unambiginity and (c) universality in naming. The International Code of Botanical Nomenclature attempts to do this by laying down certain provisions which must be followed in giving scientific names to plants. These provisions include (i) publication, (ii) typification, (iii) priority, and (iv) legitimacy.

The International Code of Botanical Nomenclature (ICBN for short) requires that names must be published in order to have any standing in the scientific naming of plants, and publication here means (i) effective publication and (ii) valid publication. For effective publication a name must be published either in books or recognized scientific periodicals. For valid publications a name must be accompanied by a description of the plant, and where applicable a reference to a previously effectively published description. Names which have not been validly published have no standing in botanical science. Since January 1935 the validating description of the name of a new taxon of higher plants must be given in Latin, so as to ensure that all descriptions are readily comparable. The author of a plant name is the person who first validly published that name.

Since January 1958 the name of a new taxon of the rank of family and below, was not considered validly published unless the nomenclature type was indicated. Typification is the process of indicating or designating "a type". 'TYPE' in botanical nomenclature is formally defined as the element on which the description validating the publication of a name is based, which in the case of a species, is the herbarium specimen from which the description was recorded. The type method fixes the application of names definitively and unambiguously. (Jeffery 1981).

The principle of priority states that when two or more names compete for the same taxon, in general, the oldest one is regarded as the correct one i.e. the first validly published one. Two or more names applied to the same taxon is known as synonyms. The principle of priority gains prominence when a species undergoes a change of taxonomic position. In such cases ICBN lays down that though a species has by virtue of new findings to be transferred to another genus, the oldest specific epithet should be retained. Thus for taxa of the rank of species and below, epithets have priority and correct name is provided by a combination of the oldest available epithet with the correct generic name. If this principle had been followed *Oxytenanthera monostigma* Bedd which was originally known as *Bambusa ritchey* Munro would have been called *Oxytenanthera ritchey.* The principles of nomenclature have not been strictly followed in bamboos.

In regard to the universality of names, ICBN lays down that in any given classification a taxon can have only one correct name. Therefore the correctness or otherwise of a name depends upon the classification adopted, together with the observance of all the requirements of the ICBN. As the classification of bamboos is made more predictive, comprehensive, and useful, there are bound to be name changes for taxonomic reasons, and many names that have been held to represent distinct species may be reduced to synonyms.

HISTORY OF BAMBOO CLASSIFICATIONS

1750—1839

The earliest attempts at describing bamboos appears to be by Rumphinus (1750) who described and named some bamboos in his publication titled "Herbarium Ambionense". Later in 1789 *Bambusa arundinacea* from India was described by Schreber. More than four decades later in 1837 Blanco described many plant species of Philippines including bamboos. But he described them all under the name 'Bambusa' and his descriptions were very short and brief. In 1839 Ruprecht published a monograph on bamboos in which he described 18 species from the Indo Malayan region. His study was based mainly on herbarium specimens.

MUNRO 1868

This work was followed by the more extensive work of Munro who published a monograph of the Bambuseae in the Trans Linean Society Journal Vol. XXVI 1868. His work was also based on herbarium specimens and he described 170 species grouped into 21 genera divided into 3 divisions. The first division consisted of genera with flowers having 3 stamens and 3 stigmas and culms without thorns eg. *Arundinaria*. The second division consisted of the so called "true bamboos" such as *Bambusa* and *Gigantochloa*. The third division consisted of genera characterised by the peculiar structure of the fruit consisting of a thick pericarp enclosing the seed eg. *Melocanna*.

KURZ 1876

In 1876 Kurz for the first time studied bamboos as living specimens in the field and recognized the importance of vegetative characters such as culm sheaths. He however did not propose any formal classification of bamboos.

GAMBLE 1881

In 1881 Gamble studied the bamboos of India, Burma and Malaya and published a monograph on Bambuseae of British India which still remains the fundamental work on Indian Bamboos. He recognized 14 genera and his work contains a comprehensive systematic account of all the species known at the time of its publication totalling 151 species. He divided bamboos into 4 subtribes viz (i) Arundinareae (ii) Eubambuseae (iii) Dendrocalameae (iv) Melocanneae as follows:

Subtribe I: ***Arundinareae* has two genera—**

(a) *Arundinaria* Mich with culms erect occasionally climbing and shrubby found with very few exceptions in the hill regions. There are 28 species altogether, described in the Flora of British India.

(b) *Phyllostachys* Sieb & Zucc culms resemble Arundinaria but the internodes are more or less flattened on one side. Two species occur in India *P.bambusoides* Sieb & Zucc in Mishmi hills of Upper Assam, *P.manii* Gamble in Khasia hills at 5000 feet altitude and in Naga hills. It is also found in upper Burma and is called Maipangpuk.

Subtribe II: *Eubambuseae* has four genera—

(a) *Bambusa* Schreb. This genus contains some of the most important species and some of the species with the largest culms. There are 22 species found in India, Burma and Ceylon. Some of them are difficult to distinguish when not in flower, and even the culm sheaths which are usually the best means of identification are sometimes difficult to recognize.

(b) *Thyrsostachys* Gamble. There are two species in this genus both of which have erect tufted culms. *T. oliveri* Gamble is found in the hills of upper Burma in moist forests on ridges at 2000 ft. altitude and also in Shan hills. It has straight culms upto 50 to 80 ft in height and 1½ to 2½ inches in diameters. The sheaths remain persistent on the culm for long and nodes are only very slightly thickened. The cavity of the internodes has a diameter of half that of the culm. *T. siamensis* Gamble is found in Burma from Mandalay down to Tenasserim. Culms are 25 to 30 feet in height 1½to 3 inches in diameter usually covered with persistent bases of sheaths and having nodes *not* prominent. The cavity is more than half the diameter of the culm.

(c) *Gigantochloa* Kurz has two species. *G. verticillata* Munro is a Malay species of very large size culms attaining 80 to 100 feet in height and 4 to 5 inches in diameter, greyish green in colour. The young culms are striped with yellow and is found wild in Tenasserim and cultivated in Calcutta Botanic Gardens. The Malay species *G. atter* Munro and *G. apus* Kurz are both found in Calcutta gardens. *G. macrostachya* is found in tropical forests in Garo hills, Assam, Chittagong, Arkansas and Burma. *G. kurzii* is a little known bamboo of Tenasserim and Malay found near the coast.

(d) *Oxytenanthera* Munro has seven species arborescent or scandent bamboos with stout usually creeping stoloniferous rootstocks. The seven species are *O.nigrociliata* Munro, *O. albociliata* Munro, *O.parvifolia* Brandis, *O. thwaitessi* Munro, *O. monostigma* Bedd, *O. stocksii* Munro, *O. Bourdilloni* Gamble.

Subtribe III: *Dendrocalameae* has five genera—

(a) *Dendrocalamus* has 15 species

(b) *Melocalamus* Benth has one species

M. compactiflorus Benth also known as *Pseudostachyum Compactiflorum* is a climbing bamboo of Sylhet, Chittagong.

(c) *Pseudostachyum* Munro *P. polymorphum* is a thin walled shrubby bamboo of river banks and valleys on terai and lower hills of Sikkim rising to 3000 ft altitude.

(d) *Teinostachyum* Munro has five species shrubby or arborescent, erect, straggling or climbing. 3 of the species are found in Assam and Burma one in S. India and one in Ceylon. *T. griffithii* Munro; *T. wightii* Bedd, *T. attenuatum* Munro, *T. dullooa* Gamble, *T. helferi* Gamble.

(e) *Cephalostachyum* Munro has seven species shrubby with spike in globose or panicled heads or in fascicles, found in eastern Himalayas, Assam and Burma. One species extends to the peninsula to the forests of Chotanagpur.

Subtribe IV: *Melocannae* contain three genera—

(a) *Dinochloa* has two species viz. *D. tjankorreh* Buse and *D. maclelandii* Kurz found in the Andaman islands and the other, in Chittagong and Burma.

(b) *Melocanna* Trin has to species of arborescent bamboos both of Burmese region, with one extending to Chittagong and Assam viz *M. bambusoides* Trin and *M. humilis* Kurz.

(c) *Ochlandra* Thiv has six species of shrubby. gregarious, reed like bamboos all of south India or Ceylon.

BENTHAM 1883

In 1883 Bentham adopted Munro's system of classification as the basis of his revision of bamboo classification. His work published in Genera Plantarum vol. III is regarded as a work of remarkable completeness considering the chaotic condition in which the author found the order. He described 18 genera of which 15 belonged to the Indo Malayan region and the rest chiefly to south America. Bentham's scheme as quoted by Holttum (1956) is as follows:-

Bentham divided bamboos into four subtribes.

Subtribe I: *Arundinariae*

stamens usually 3, palea 2 keeled, pericarp thin, and adnate to the seed. Contains 3 genera *Arundinaria, Phyllostachys,* and *Chusquea*.

Subtribe II: *Eubambuseae*

Stamens 6, palea usually 2 keeled, pericarp thin, adnate to the seed.

Contains 6 genera *Nastus, Guadua, Bambusa, Thyrsostachys, Gigantochloa* and *Oxytenanthera.*

Subtribe III: *Dendrocalameae*

stamens 6, palea 2 keeled, pericarp fleshy or crustaceous, separable from the seed. Contains five genera *Dendrocalamus, Melocalamus, Pseudostachyum, Teinostachyum,* and *Cephalostachyum.*

Subtribe IV: *Melocanneae*

Stamens 6 or more, spikelet one flowered, palea none or similar to the flowering glume, pericarp crustaceous or fleshy separable from the seed. Contains four genera *Dinochloa, Schizostachyum, Melocanna, Ochlandra.*

STAPF 1897

In 1897 Stapf classified bamboos into five subtribes and regarded bamboos as a tribe of the subfamily Poideae of the family Gramineae. His classification is based on the character of the fruit, nature of spikelets and paleas, and the number of stamens as given below:-

Subtribe I: *Dendrocalameae*

Fruit is a caryopsis, stamens 6 in number, paleas 2 keeled, spikelets one to many flowered. Genera *Dendrocalamus, Melocalamus, Pseudostachyum, Teinostachyum* and *Cephalostachyum.*

Subtribe II: *Melocanneae*

Fruit is a berry or nut, stamens 6 in number and paleas 2 keeled Genera *Dinochloa, Scizostachyum, Melocanna* and *Ochlandra.*

Subtribe III: *Bambusineae*

Stamens 6 in number and upper florests imperfect. Genera *Bambusa, Gigantochloa, Oxytenanthera, Oreobamboos, Guadua* and *Nastus.*

Subtribe IV: *Arundinareae*

Stamens 3 in number rarely 6, and lower florests perfcet. Genera *Arundinaria, Sasa, Phyllostachys, Chusquea, Arthrostylidium,* and *Merostachys.*

Subtribe V: *Puellineae*

All genera are herbaceous, stamens 6 in number, lower florets usually male, female or barren. Genera *Puellia, Gauduella,* and *Atractocarpus.*

HOOKER 1897

Hooker (1897) in the classification of the Indian genera of Gramineae, diverges somewhat from Bentham's arrangement and abandons some of the tribes and subtribes and in many cases follow Hackel's (1889) views. In the description and revision of some of the very difficult genera, Hooker has also taken the help of Stapf. Hooker considers bambuseae as a tribe of the family Gramineae and divides them into four subtribes as follows:-

Tribe XI Bambuseae — Shrubby or arboreus grasses, leaves flat, jointed on the sheath. Spikelets one to many flowered, lower two or more glumes empty, gradually increasing in size upto the flowering, with sometimes small terminal imperfect ones. Palea usually large 2—keeled (in Melocanna). Lodicules usually three, stamens 3, 6 or many. Styles 2 or 3.

Subtribe I: ***Arundinareae***

Has two genera *Arundinaria* and *Phyllostachys.*

Subtribe II: ***Eubambuseae***

Has four genera *Bambusa, Thyrsostachys, Gigantochloa,* and *Oxytenanthera.*

Subtribe III: ***Dendrocalameae***

Has five genera *Dendrocalamus, Melocalamus, Pseudostachyum, Teinostachyum* and *Cephalostachyum.*

Subtribe IV: Melocannae has four genera *Schizostachyum, Dinochloa, Melocanna,* and *Ochlandra.*

CAMUS 1913

In 1913 bamboos of Indo-china was described by Camus, and he adopted Munro's system of classification and published his work titled "Les Bambusees". He treats bamboos as a subfamily of the Gramineae and divided them into five tribes and four subtribes. Engler and Prantl (1915) followed the classification of Hackel in Die Naturlichen Pflanzenfamilien.

BRANDIS 1921

In 1921 Brandis developed a key for the identification of 14 genera of bamboos based on culm habit, size of culms, leaf characteristics, floral characteristics and fruit and seed characteristics as follows:-

I. Culms as a rule over 20 feet (6 meters) in height, stamens usually 3, pericarp thin, membranous, and adnate to the seed.

A. Branches terete, transverse veins conspicuous in most species dividing the leaves into rectangles or squares. Spikelets often pedicelled 1 to many flowered, empty glumes 1 to 2 ______________ ***Arundinarea***

B. Branches flattened on the inside transverse veins always conspicuous close together, usually dividing the leaf into minute squares. Spikelets sessile, supported by prominent sheathing bracts, often with a leafy blade, flowers 1 to 4 empty glumes 2 to 3 ______________ ***Phyllostachys***

II. Culms tall as a rule erect, stamens 6, pericarp thin, membranous, adnate to the seed.

A. Filaments free.

(a) Paleae entire or slightly 2-dentate all prominently 2-keeled

——————***Bambusa***

(b) Paleae deeply 2-dentate, teeth awned, the uppermost nearly entire indistinctly keeled

——————***Thyrsostachys***

B. Filaments Connate

(a) Spikelets many flowered, paleae all prominently keeled

——————***Gigantochloa***

(b) Spikelets few flowered, paleae of upper flowers indistinctly or not at all keeled ——————***Oxytenanthera***

III Culms tall sometimes climbing, stamens 6, in Ochlandra numerous.

A. Fruit small, pericarp crustaceous, endosperm large.

(a) Single stemmed, culms overhanging, transverse veins conspicuous ——————***Pseudostachyum***

(b) Tufted culms, transverse veins as a rule not conspicuous.

(i) Lodicules none, spikelets 2-6, flowers in large globose heads.

——————***Dendrocalamus***

(ii) Lodicules 3, conspicuous.

1. Spikelets in long narrow spikes. Spikelets 2 to 5 flowered.

—————— ***Teinostachyum***

Spikelets one flowered —————— ***Schizostachyum***

2. Spikelets crowded in globose heads or obconical heads.

——————***Cephalostachyum***

B. Fruits large, pericarp fleshy or ultimately coriaceous, no endosperm in the ripe seed.

(a) Stamens 6, sometimes 4 or 5.

(i) Lodicules none, climbing culms zigzag, geniculate, spikelets one flowered and minute ——————***Dinochloa***

(ii) Lodicules 2 or 3

1. Loosely tufted culms spreading often climbing nearby trees. spikelets small in distant compact globose heads. ——————***Melocalamus***

2. Usually single stemmed spikelets in large panicles ——————***Melocanna***

(b) Stamens 6 to 120, stamens often overhanging ——————————***Ochlandra.***

BACKER 1924. AGNES ARBER 1926

Backer (1924) described some bamboos while writing the flora of Java and many of Rumphinus names were mentioned as synonyms. Agnes Arber (1926) made a comparative study of the spikelets of twenty species belonging to eight genera of bamboos mostly from herbarium material by cutting serial microtome sections. The classification favoured by her is that of Bentham's Genera Plantarum. She found that the entire shoot apex is transformed into an axillant leaf and no tissue whatsoever is left over to form the rudiment of a growing point in *Oxytenanthera nigrociliata* Munro and *Gigantochloa scortechinii* Gamble. Further, she also found that the commonest type of vascular system in the bamboo ovary is the one in which a dorsal strand supplies the ovule, and two others lateral and posterior run up the ovary wall. This arrangement was observed in *Gigantochloa latispiculata* Gamble, *Oxytenanthera nigrociliata* Munro, *Dendrocalamus sikkimensis* Gamble, *Dendrocalamus strictus* Nees, *Dendrocalamus giganteus* Munro and *Schizostachyum brachycladium* Kurz. The placental strand does not arise above the ovule, but the three strands pass into the base of the style. In species with three stigmas eg. *Bambusa nutans* Wall and *Schizostachyum brachycladium* Kurz each strand passes into one of the stigmas.

The orthodox view taken by Hackel is to regard grass gynaecium as monocarpellary. But the occurrence of the type of vascular symmetry described by Agnes Arber led Schuster to return to the view of Celakosky et al that gynaecium in grasses is tricarpellary, the three bundles being midribs of the carpels, and the single ovule being borne on the suture between the two posterior carpels.

CAMUS 1935. HOLTTUM 1946-1956

In 1935 Camus proposed a new scheme in which all genera having filaments united to form a tube were placed in a separate subtribe. This arrangement was challenged by Holttum who published a note on it in 1946. In the same note he also pointed out that the distinctions between Bentham's subtribes based on fruit structure did not correspond with facts, and he suggested an alternative scheme covering the genera of the Malayan region. Holttums work (1946) is now basic to Malaya and Indonesia.

Holttum considered Bentham's classification unsatisfactory and in need of complete revision because he felt that both Munro and Bentham had examined very few fruits, and their limited observations were insufficient for the important place given to fruits in the scheme. Holttum examined fresh fruits of *Bambusa* and *Gigantochloa* and found them not essentially different in structure from those of *Dendrocalamus* though the latter is placed in a different subtribe by Bentham. Secondly *Dendrocalamus* and *Gigantochloa* was observed by Holttum to be so similar in spikelet structure that it was difficult to define the difference between them clearly. Further, he found that the three genera *Pseudostachyum, Teinostachyum* and *Cephalostachyum* of Bentham's subtribe 3 were very similar to *Schizostachyum* placed in Bentham's subtribe 4. He also felt that *Dinochloa* which was placed along with *Schizostachyum* by Bentham should be transferred and placed along with *Bambusa* in subtribe 2.

Holttum suggested a classification of bamboos based on the structure of the ovary in 1956. He recognized four types of ovary in bamboos viz (A) *Schizostachyum* type of ovary found in *Melocanna, Ochlandra* and *Schizostachyum;* (B) *Oxytenanthera* type of ovary found in *Oxytenanthera;* (C) Bambusa-Dendrocalamus type of ovary found in *Melocalamus, Dinochloa, Thyrsostachys, Bambusa, Guadua, Dinochloa, Dendrocalamus, Gigantochloa, Racemobambos,* and *Nastus.* (D) Arundinaria type of ovary found in *Arundinaria.*

The scheme of classification suggested by Holttum (1956) is as follows:-

A. Schizostachyum type of ovary, pericarp thick and and fleshy throughout; one flower in spikelet, stamens 6, occurring in Burma region.
——— ***Melocanna.***

Stamens more than 6 occurring in South India and Ceylon.
——— ***Ochlandra.***

Pericarp thin and dry thickened slightly at the apex only; one to many flowers in spikelet.
Schizostachyum
Teinostachyum
Pseudostachyum
Cephalostachyum
Neuhouzeaua.

B. Oxytenanthera type of ovary.
——— ***Oxytenanthera.***

C. Bambusa - Dendrocalamus type of ovary.
Inflorescence of spikelet tufts. Pericarp of mature fruit uniformly thick and fleshy. spikelets with 2 or 3 perfect flowers and a rudiment.
——— ***Melocalamus.***

Spikelet with one perfect flower and no rudiment.
——— ***Dinochloa.***

Pericarp of mature fruit slightly fleshy at apex only. Spikelet usually many flowered always with a rudiment at the apex. Internodes of rachilla elongate and jointed below each lemma. Lemmas all equal. Lower paleas deeply bilobed.
——— ***Thyrsostachys.***

Lower paleas not deeply bilobed occurring in tropical Asia.
——— ***Bambusa.***

Occurring in tropical America.
——— ***Guadua.***

Spikelets 1 to 6 flowered, apical rudiment present or absent, rachilla internodes always very short and not jointed below lemmas. If several flowers, lower lemmas shorter than upper, style short, stigmas 3, one flower and no rudiment beyond it.

——————————— ***Dinochloa.***

Style long, stigma usually-1, uppermost flower perfect, or various types of rudiments sometimes present; stamen tubes sometimes present, always very delicate.

——————————— ***Dendrocalamus.***

Uppermost flower always consisting of a narrow long empty lemma. A rather firm and very distinct stamen tube always present.

——————————— ***Gigantochloa.***

Inflorescence a raceme or panicle

Spikelets many flowered. ——————————— ***Racemobambos.***

Spikelets one flowered. ——————————— ***Nastus incl Chloothamnus.***

D. Arundinaria type of ovary, top of ovary not thickened.

——————————— ***Arundinaria.***

RAIZADA and CHATERJEE 1963

Seven years later in 1963, Raizada and Chaterjee working in the Forest Research Institute at Dehradun, attempted to differentiate bamboo species based on culm sheaths, as they observed that the culm sheaths are more or less distinctive for individual species of bamboos, and the genera *Bambusa* and *Dendrocalamus* have special characteristics. They found that the general appearance, size, texture and shape of culm sheaths and their blades are particularly useful in distinguishing species. Their observations were based on 22 species growing in the Demonstration Area of New Forest—the campus of the Forest Research Institute at Dehradun.

They observed that the collosal size (35.5 to 60.9 cm in width at the base) of the culm sheaths in *Dendrocalamus giganteus, D. calostachyus* and *D. brandisii* separates them from the rest of the species. Further, the cylindrical type of sheaths are characteristic of *Dendrocalamus longispathus* and *D. membranaceous.* On the other hand in *Bambusa nutans, B. tulda* etc the sheaths are oblique and asymmetrical, while in *B. pallida, D. strictus,* and *D. hamiltonii* the sheaths are conical in shape. A singularly distinguishing character according to them is the brown or black felted blade in *B. arundinacea.* Then again in *Melocanna bambusoides* the blade is subulate and pointed with a very prominent auricle.

They also observed that sometimes the sheaths of different species are very much alike and within the species itself there is a large variation in size, shape etc. For instance the sheaths of *Bambusa tulda, B. nutans* and *B. burmanica* are very much alike, but auricles in *B. tulda* are quite distinct from the blade, and very much laterally placed on top of the sheath. In *B. nutans* on the other hand auricles are more or less continuous with the blade and not so much laterally placed. The sheaths of *B. tulda* and *B. burmanica* they found are difficult to differentiate, and in both there are two white bands one on each side of the node. They also found that there is a great variation in size of sheaths in *Dendrocalamus hamilitonii* where both big sized and small sized ones occur, though the smaller ones are more common. Those of the larger types were observed to be 40.6 to 45.7 cm long and 20.3 to 40.6 cm broad at the base, while the smaller ones were 25.4 to 35.5 cm long and 12.7 to 20.3 cm broad at the base. Further in *Dendrocalamus membranaceous* the sheaths are both cylindrical and domed, though the cylindrical type is more common. Then again in *Bambusa arundinacea* sheaths are both cylindrical and broadly triangular in shape though the former is more common. In *Bambusa vulgaris* also the sheaths are both big and small.

Based on these findings, Raizada and Chaterjee developed a key for the identification of the 22 species studied by them. But the deliniation of species throughout the key is very difficult of delimitation and definition.

FUJIMOTTO 1966 and MCLURE 1996

In 1966, Fujimotto attempted a classification of Bambusoideae based on the leaf structure especially the ligule portion, and his approach was purely phenetic. Another significant work on bamboos appeared the same year titled "Bamboos - a Fresh Perspective" by McClure. He pointed out that all parts of the vegetative characters, and the flower structure, should be used for bamboo classification.

The nineteen sixties was a period when a number of independent studies were being carried out in different parts of the world on different aspects of bamboos. Among the more significant ones of these studies are the anatomical studies of the culms as an aid to identification and classification of bamboos carried out (i) by Pattanath and Rao working in the Forest Research institute at Dehradun India, (ii) Grosser and Zamuco on Philippine bamboos, and (iii) Grosser and Leise working in Germany on bamboos obtained from India, Pakistan, Indonesia, Thailand, Philippines, Taiwan and Japan.

GHOSH and NEGI 1960

Studies carried out by Pattanath and Rao were in fact an extension of studies carried out by Ghosh and Negi 1960 who concentrated mostly on the gross features of the epidermis of the culm of a limited number of specimens of six species of bamboos of the genera *Bambusa*, *Dendrocalamus* and *Melocanna*. Ghosh and Negi found that the epidermis of bamboo culms were remarkably uniform, consisting of long cells alternating with pairs of short cells. The middle lamella of long cells, they found, was strongly suberized and varied in wall thickness— the outer wall being thick with cuticle, and inner comparatively thin and profusely pitted. End walls of long cells were found to be straight, oblique, concave or convex. The length of long cells were found to be variable, but the width showed little variation. Short cells were reported to be arranged in pairs alternating with long cells and varying from 1 to 3 pairs. Stomata or microchairs sometimes replaced short cells. Different types of hairs such as spines and bicellular hairs were reported to occur in different species. They however laid emphasis on quantitative characters such as dimensions of different types of cells, and number of pairs of short cells, stomata and hairs per unit area etc, rather than qualitative characters for differentiation of species.

PATTANATH 1965. PATTANATH and RAO 1969

A more detail study was carried out by Pattanath (1965) and Pattanath and Rao (1969) on seventeen species of bamboos belonging to eleven genera viz (1) *Bambusa arundinacea* Willd (ii) *B. nutans,* Wall (iii) *B. polymorpha,* Munro (iv) *B. tulda,* Roxb (v) *Cephalostachyum pergracile,* Munro (vi) *Dendrocalamus hamiltonii* Nees et Arn (vii) *D. longispathus,* Kurz (viii) *D. strictus,* Nees (ix) *Melocanna bambusoides* Trin (x) *Oxytenanthera abyssincia (A. Rich)* Munro (xi) *O. nigrociliata* Munro (xii) *thyrsostachys oliveri,* Gamble (xiii) *Ochlandra travancorica,* Benth (xiv) *Teinostachyum dullooa,* Gamble (xv) *Pseudostachyum polymorphum* Munro (xvi) *Dinochloa maclelandii* Kurz (xvii) *Gigantochloa macrostachya,* Kurz.

The first twelve species were studied in considerable detail taking into consideration topographical variation within individual culms, culm to culm variation within a clump, and culm to culm variation in culms of different clumps. The specimens for these studies were obtained from the Demonstration Areas of New Forest in Dehradun. The rest of the five species were studied from limited number of specimens obtained from the State Forest Departments and the Forest Research Institute Museum.

The study brought out a number of interesting trends viz (i) the epidermis of the culm is the most stable character in bamboos exhibiting very little variation both topographically and from culm to culm confirming the earlier preliminary finding of Ghosh and Negi (1960) (ii) The epidermis as seen in surface view consists of long cells alternating with pairs of short cells which are replaced here and there with stomata or different types of micro and macro hairs again confirming the finding of Ghosh and Negi 1960, (iii) The shape of long cells, the nature of their walls and arrangement of papillae on the long cells are important features of diagnostic value. Pattanath and Rao (1969) differed from Ghosh and Negi in the importance given to the end walls of long cells by the latter, as a diagnostic feature. Further Ghosh and Negi had failed to observe the papillae on long cells reported by Pattanath (1965) and Pattanath and Rao (1969) as a general feature of bamboo epidermis. The arrangement of papillae on long cells were observed to vary from species to species providing an important diagnostic character (Pl. 2 figs 1 to 8); (iv) The structure of the stomata and micro hairs were also observed to be very distinctive characters of diagnostic value (Pl. 12 Figs. 5 to 14);

(v) Pattanath and Rao found that these qualitative characters were more or less stable, and therefore of more diagnostic value than the quantitative characters suggested by Ghosh and Negi such as the size of long cells, number of pairs of short cells, stomata and micro-hairs per unit area. Pattanath and Rao found these quantitative characters to be of limited value, as they were found to vary topographically as well as from culm to culm;

(vi) Cross section of the internodes of the culm show certain characters which are found to be common in all bamboos. The exterior of the culm is bound by an epidermis, the cells of which are generally very thick walled. Below this, is usually found a narrow zone or cortex separating the epidermis from the vascular region. In a cross section the fibro-vascular region especially of internodes at the bottom levels of the culm vary markedly from species to species and are of some diagnostic value (Pl. 3a & 3b).

(vii) The cellular structure of the cortex differs in different species and may be made up entirely of thin walled irregularly shaped closely fitting parenchyma cells where no intercellular spaces are present eg. *Bambusa nutans,* or they may be made

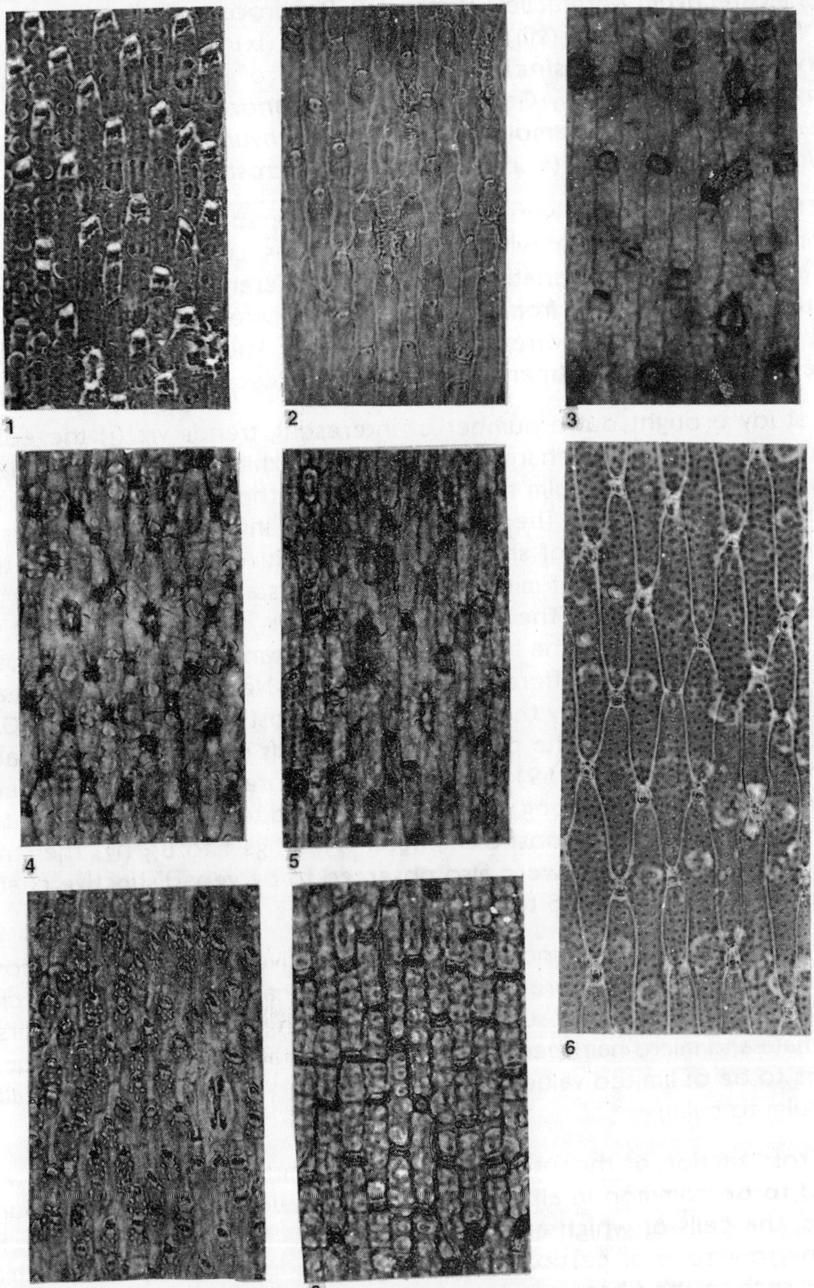

Pl. 2. Surface view of epidermis of the culm in different bamboo species (Magnification 400X). 1. *Dendrocalamus longispathus* 2. D. *Strictus* 3. D. *hamiltonii* 4. Bambusa polymorpha 5. B. *tulda* 6. *Thyrsostachys oliveri* 7. *Bambusa arundinacea* 8. *Melocanna Bambusoides*.

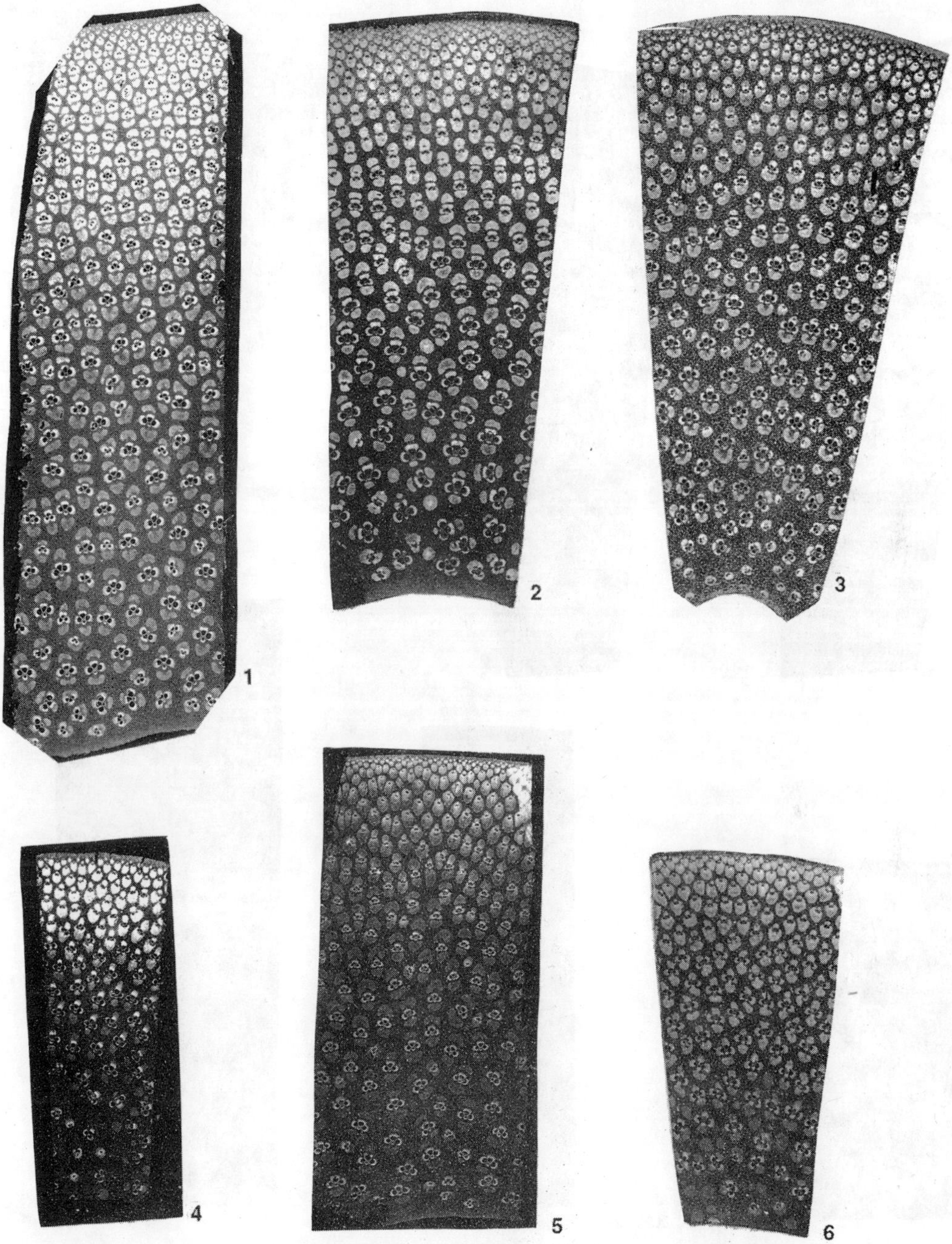

Pl. 3a. Cross sectional view of the internodes at bottom levels of the culm in some species of bamboos (Magnification 5X) 1. *Dendrocalamus hamiltonii* 2. *D. longispathus* 3. *D. strictus* 4. *Oxytenanthera abyssincia* 5. *O. nigrociliata* 6. *Melocanna bambusoides*.

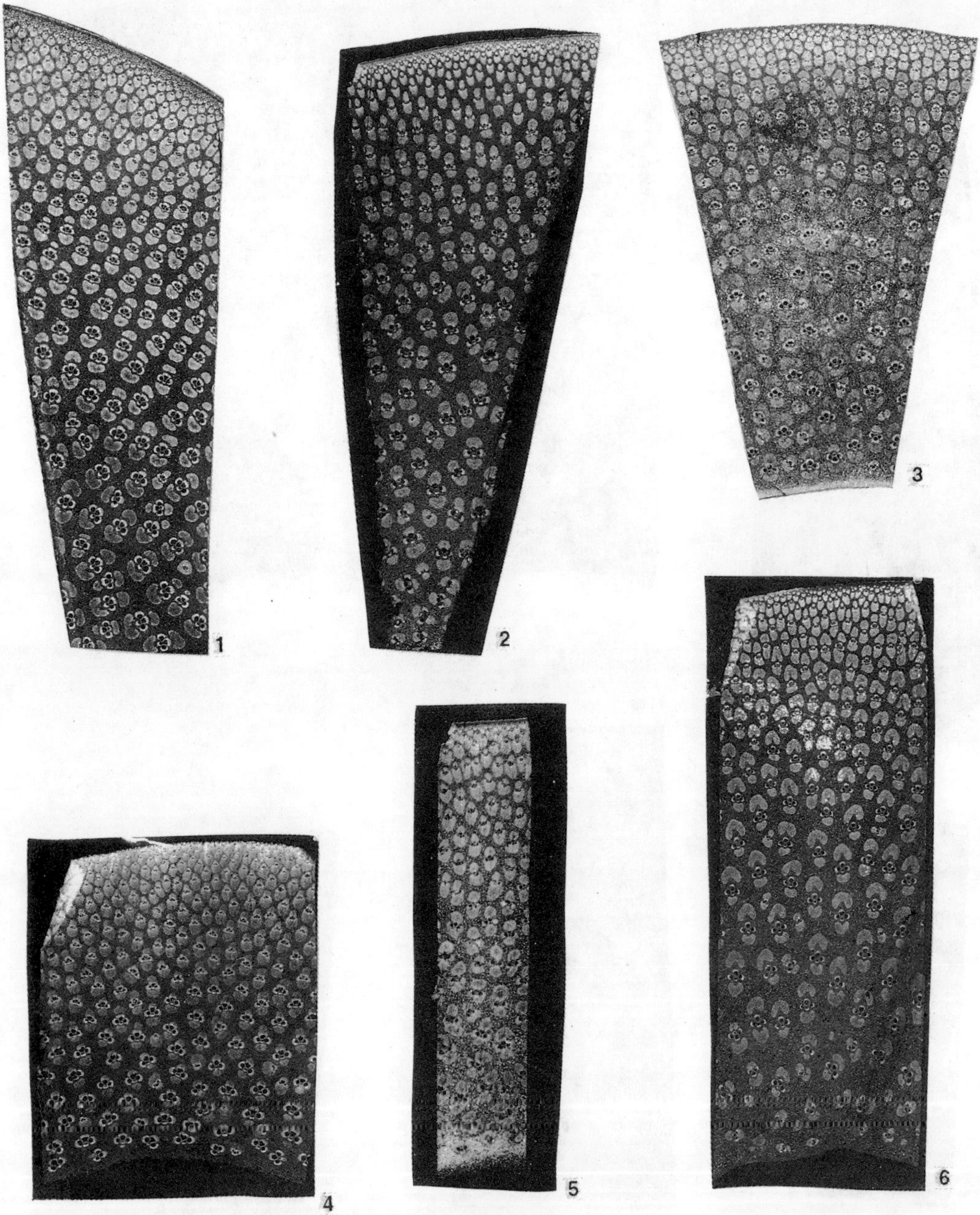

Pl. 3b. Cross sectional view of the internodes at bottom levels of the culm in some species of bamboos (Magnification 5X) 1. *Bambusa tulda* 2. *B. polymorpha* 3. *B. arundinacea* 4. *B. nutans* 5. *Cephalostachyum pergracile* 6. *Thyrsostachys oliveri*.

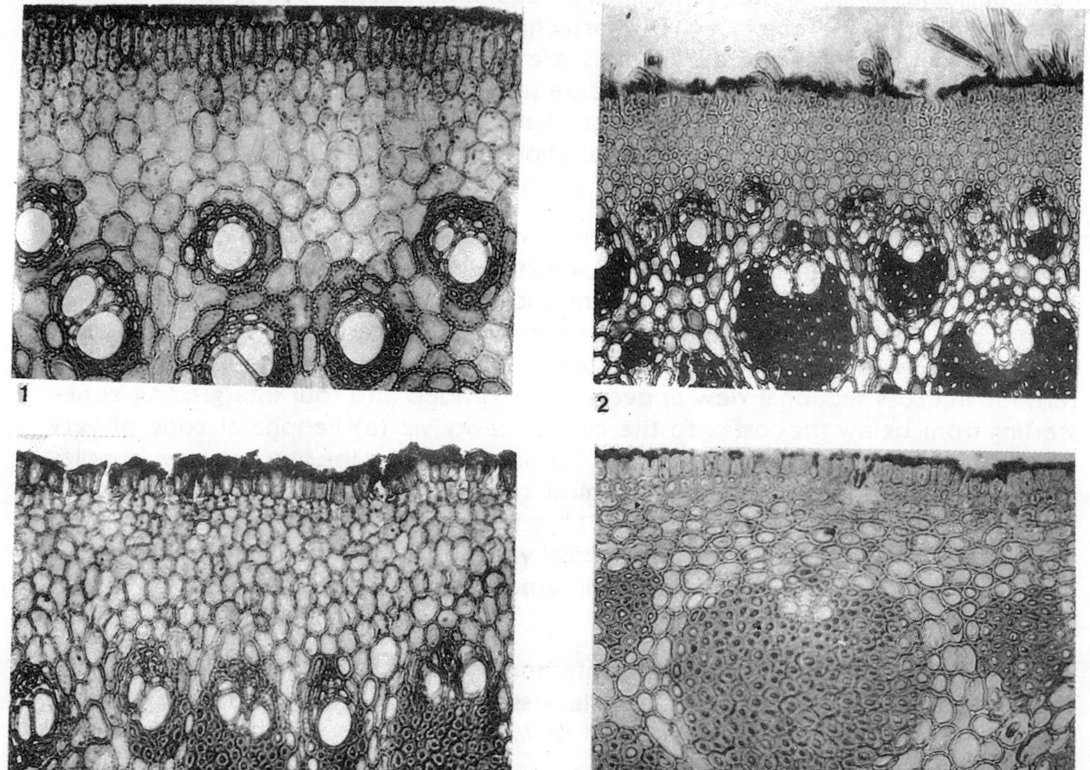

Pl. 4. Diagnostic features of the cortex of the internodes in some species of bamboos (Magnification 200X) 1. *Bambusa tulda* 2. *B. polymorpha* 3. *Cephalostachyum pergracile* 4. *Melocanna bambusoides* 5. *Oxytenanthera nigrociliata* 6. *O. abyssinica*.

up of thick walled parenchyma cells with intercellular spaces between them eg. *Dendrocalamus longispathus.* Sometimes groups of fibres and thickwalled elongated cells are found in the cortex eg. *Thyrsostachys oliveri,* and *Bambusa tulda.* In some species, groups of tracheid like cells are found scattered in the cortex eg. *Oxytenanthera abyssinica.* In some species, a subcortical layer separates the epidermis from the cortex eg. *Bambusa tulda.* These characters are stable and show hardly any topographical variation, and very little variation from culm to culm, and are therefore of diagnostic value for differentiation of species. (pl. 4)

viii) Below the cortical region is the fibro-vascular region which in cross sectional view exhibit a highly variable appearance within an internode from epidermis to central cavity, as well as topographically from internode to internode from bottom to top of the culm. The pattern of variation however was observed to be stable from culm to culm and therefore could be characteristic of the species. (Pl 3a & 3b). Each internode in cross sectional view appears to be divided into four intergrading zones starting from below the cortex to the central cavity viz (a) Peripheral zone of very reduced fibro-vascular bundles; (b) Transitional zone where the fibro-vascular bundles are a little more differentiated; (c) Central zone where the fibro-vascular bundles are fully and completely differentiated; (d) Inner zone where the fibro-vascular bundles again undergo reduction. The topographical variation is restricted to the central and inner zone vascular bundles within the internodes, starting from the bottom internode to the top internodes. (Pl. 6)

(ix) In the peripheral zone of any internode the fibro-vascular bundles are very reduced and generally consists of a single vessel completely surrounded by fibres or partially surrounded by fibres. (Pl. 5 fig 1 to 7)

(x) In the transitional zone the vascular bundles are more developed and consist of two metaxylem vessels placed side by side, one protoxylem vessel below, and a phloem group above, all arranged in the form of a cross and surrounded by a fibre sheath which is thicker towards the protoxylem end. The shape of the vascular bundles in the transitional zone may be rounded, oval, bottle-shaped, or shaped like number eight. The shapes of the fibro-vascular bundles in the transitional zone as seen in cross sectional view, show little topographic variation, but vary from species to species and may therefore be used as a diagnostic character. (Pl 5 fig 8 to 12)

(xi) In the central and inner zones, the fibro-vascular bundles are more fully differentiated and consist of the same xylem and phloem elements. In these two zones, the four conducting elements are each surrounded by independent crescent shaped sheaths of fibres. Further, in addition to the sheaths, are the fibre caps which vary from one to four in number but generally two. When four in number, they are attached to their respective sheaths as in *Cephalostachyum pergracile.* In general the fibre caps consist of fibres which are broader and longer than the sheath fibres. When only two fibre caps are present, one is above the phloem sheath, and the other below the protoxylem sheath. These caps may be separated from their respective sheaths by parenchyma cells as in *Thyrsostachys oliveri,* or the phloem fibre cap may be fused with the phloem fibre sheath as in *Bambusa polymorpha.* These fibre caps may be sagitate, cordate, lunate, reniform or crescent shaped in outline. The relative sizes of the caps and sheaths also vary from species to species and to a certain extent within individual culms. (Pl 3a & 3b, Pl 5 fig 13 to 17)

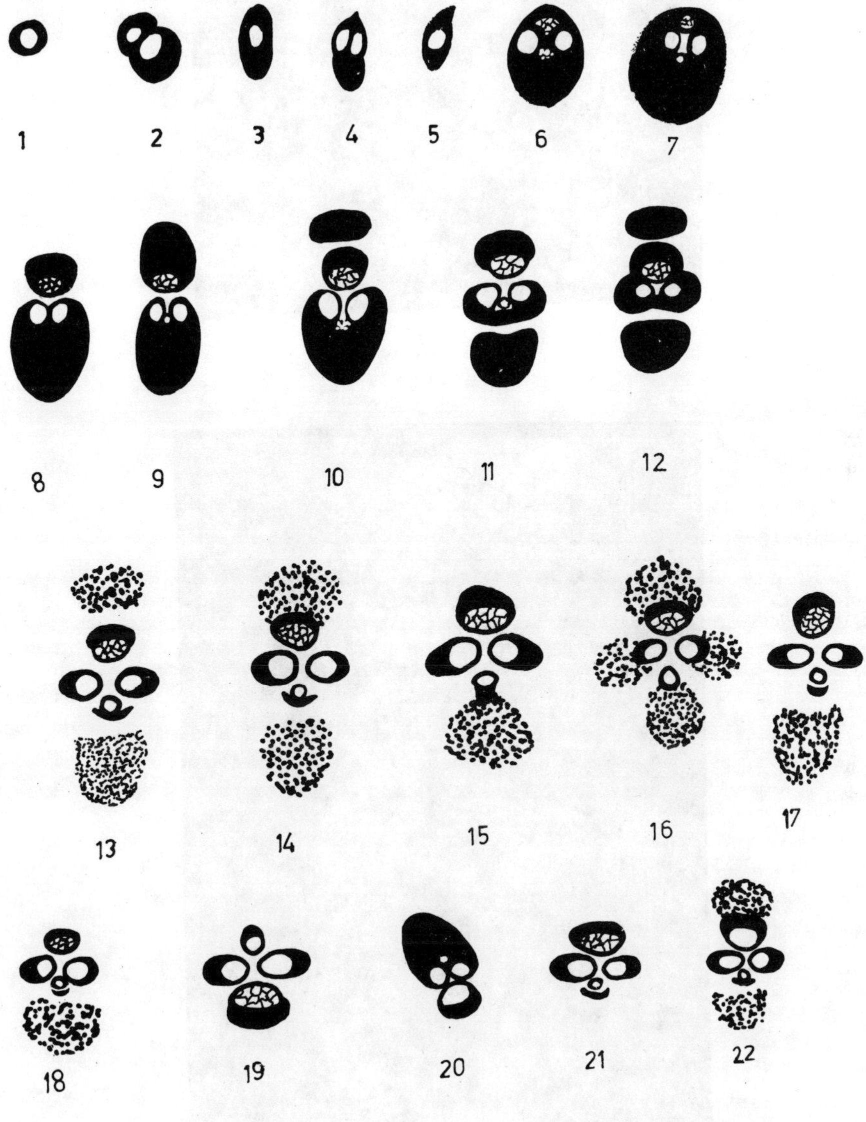

Pl. 5. Types of fibro vascular bundles in four different zones as seen in cross sections of internodes at bottom levels of the culm. 1 to 7. Peripheral zone. 8 to 12. Transitional zone. 13 to 17. Central zone. 18 to 22. Inner zone.

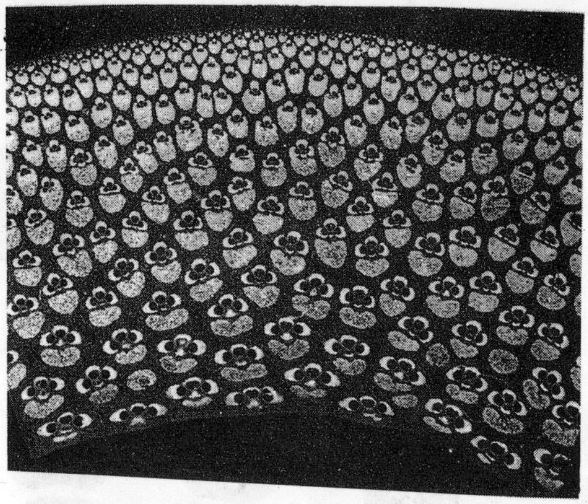

2

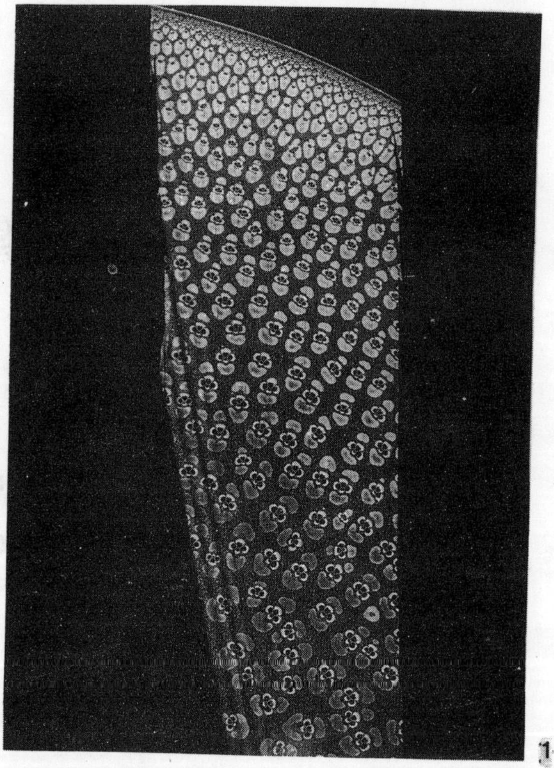

1

Pl. 6. Variation in fibro-vascular bundles as seen in cross sections of internodes at two different levels of the culm of *Bambusa tulda*. 1. Third internode from bottom of the culm (Magnification 5X). 2. Twelveth internode from bottom of the culm (Magnification 7X).

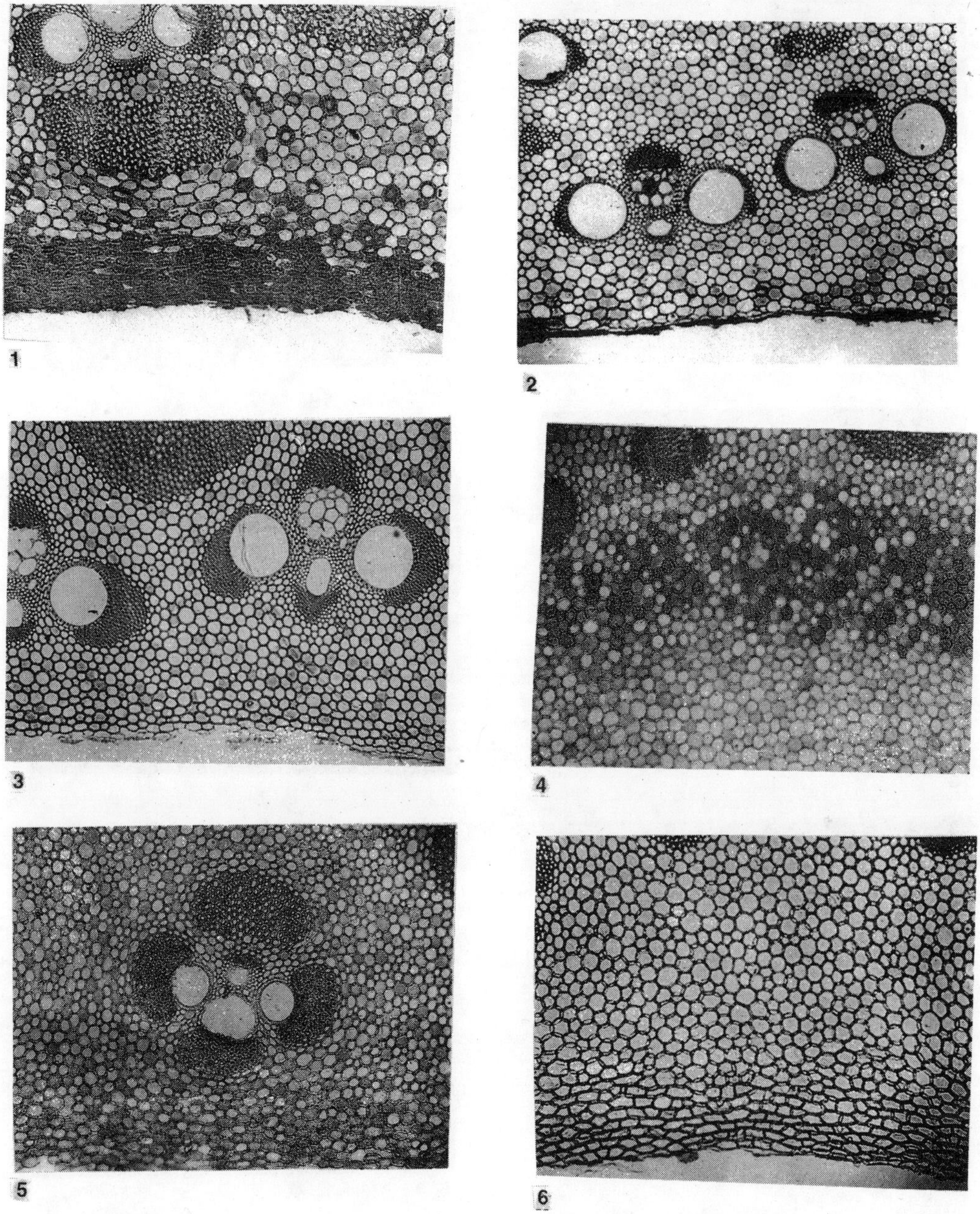

Pl. 7. Cells lining central cavity of internodes at bottom levels of the culm in different bamboo species (Magnification 60X) 1. *Dendrocalamus longispathus* 2. *Oxytenanthera abyssinica* 3. *Thyrsostachys oliveri* 4. *Cephalostachyum pergracile* 5. *Bambusa polymorpha* 6. *Dendrocalamus hamiltonil*.

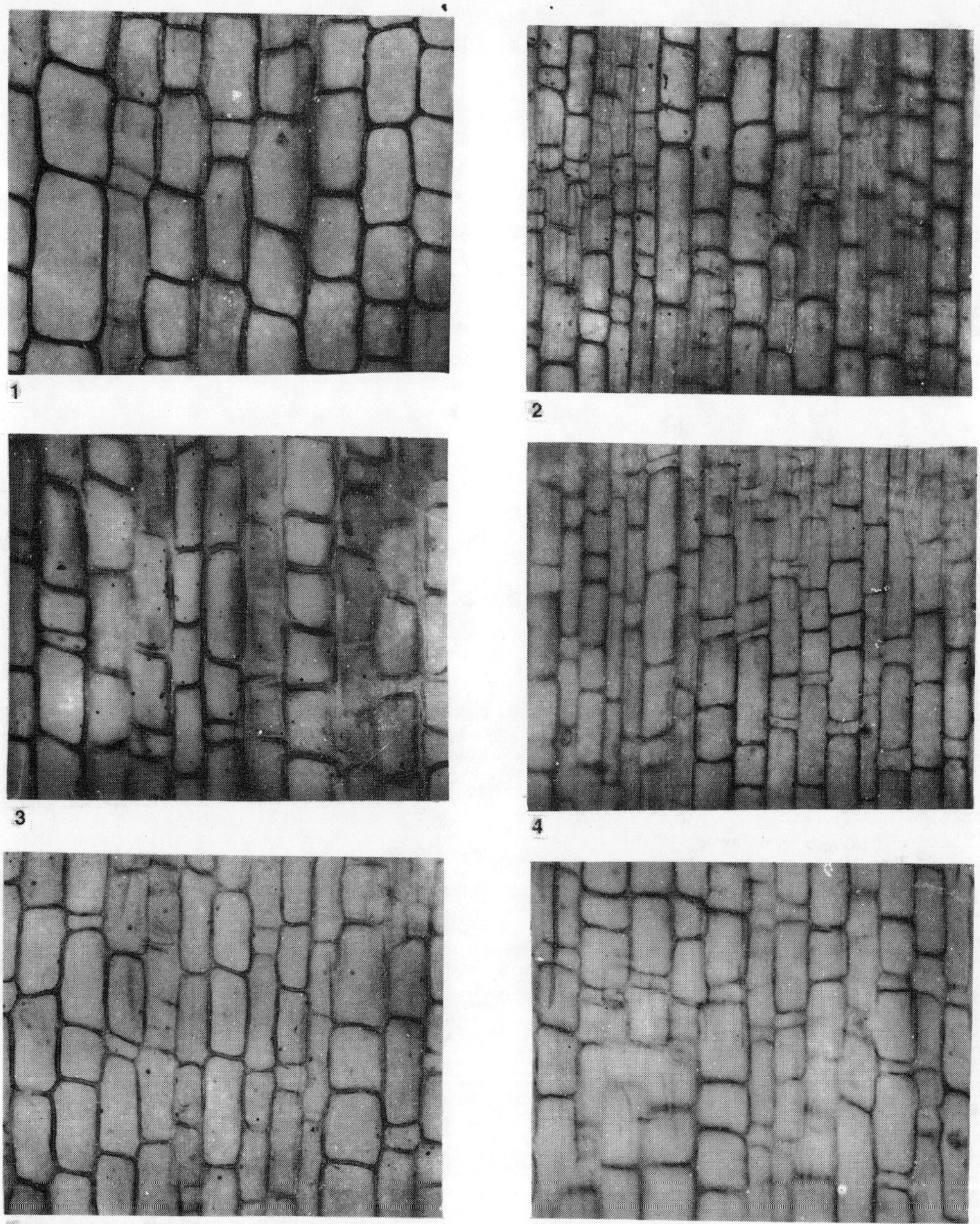

Pl. 8. Parenchyma forming ground tissue as seen in radial sections of the internodes of the culm in different species of bamboos (Magnification 200X). 1. *Dendrocalamus longispathus* 2. *Bambusa tulda* 3. *Cephalostachyum pergracile* 4. *Oxytenanthera abyssinica* 5. *Thyrsostachys oliveri* 6. *Melocanna bambusoides*.

(xii) Topographical variation was observed mostly in the fibro-vascular bundles of the central and inner zones of the internodes. Starting from the bottom internodes upwards a reduction in the number of fibre caps was observed in species which have two fibres caps in the lower internodes eg. in *Bambusa tulda* the lower internodes have both the phloem fibre cap as well as the xylem fibre cap. In the successively higher internodes however the phloem fibre cap disappears, and only the protoxylem fibre cap remains. (Pl 6.) In the inner fibro-vascular zone, the fibre caps are mostly absent and the four vessels are supported by independent sheaths. In some species there is a mixture of different types of vascular bundles in the transition zone, central zone and inner zone which is more evident in lower internodes.

(xiii) The type and shape of cells that surround the central cavity has also been observed to be of some diagnostic value, especially in the lower internodes of the culm, as in *Dendrocalamus longispathus* where a band of thick walled sclerieds surround the central cavity (Pl 7). Further, the parenchyma which form the ground tissue as seen in radial section also differes to a certain extent from species to species (Pl 8).

Pattanath (1965) listed a number of diagnostic characters for differentiation of bamboo species (Table 1), and showed that in addition to the epidermis in surface view, there are even in the cross sectional view of the internodes, a number of characters which are not so variable as the fibro-vascular bundles of the central and inner regions, and which could be used for differentiating different species of bamboos.

Pattanath and Rao (1969) did not propose any system of classification or grouping of bamboos, but they believed that the characters listed in Table 1 could be used in different combinations for identification of all species of bamboos. They also drew attention to the fact that in their studies the two genera *Bambusa* and *Dendrocalamus* which have been placed in two different subtribes by Gamble (1881) and Bentham (1883) have several anatomical features in common, the exception being *Bambusa polymorpha*. Further, they also found that differentiation into genera was not possible and there were many overlapping characters.

Table - 1 List of Diagnostic Anatomical Features of Bamboo Culms.
(Source : Pattanath 1965)

Sl. No.	Diagnostic Features	Species in which it occurs
1.	Long cells of epidermis rectangular in shape.	*Bambusa arundinacea, B. nutans, B.tulda, Dendrocalamus hamiltonii D. longispathus, D.strictus, Melocanna bambusoides, Oxytenanthera abyssinica, O.nigrociliata, Pseudostachyum polymorphum, Gigantochloa macrostachya, Dinochloa maclelandi*

2.	Long cells rhomboidal in shape.	*Bambusa polymorpha, Cephalostachyum pergracile, Thyrsostachys oliveri Ochlandra travancorica. Teinostachyum dullooa.*
3.	Average length of long cells more than 50 microns	*Bambusa polymorpha, Dendro - calamus hamiltonii, Oxytenanthera nigrociliata, Thyrsostachys oliveri, Ochlandra travancorica. Teinostachyum dullooa. Pseudostachyum polymorphum.*
4.	Lateral walls of long cells straight to slightly sinuous	*Bambusa polymorpha, Cephalosta - chyum pergracile, Dendrocalamus strictus, Ochlandra travancorica, Gigantochloa macrostachya.*
5.	Lateral walls of long cells straight	*Dendrocalamus longispathus, Oxytenanthera nigrociliata, Thyrsostachys oliveri, Teinostachyum dullooa.*
6.	Lateral walls of long cells sinuous to wavy.	*Bambusa arundinacea, B.nutans, B.tulda, Dendrocalamus hamiltonii, Melocanna bambusoides, Oxytenanthera abyssinica, Pseudostachyum polymorphum, Dinochloa maclelandii*
7.	Papillae on long cells scattered singly.	*Bambusa arundinacea, B.nutans, Dendrocalamus hamiltonii. D.strictus, Oxytenanthera abyssinica, O.nigrociliate, Ochlandra travancorica, Teinostachyum dullooa, Pseudostachyum polymorphum, Gigantochloa macrostachya. Dinochloa maclelandii*
8.	Papillae on long cells in scattered groups.	*Bambusa polymorpha, B.tulda, Cephalostachyum pergracile, Dendrocalamus longispathus, Melocanna bambusoides.*
9.	Papillae on long cells in one or two large groups.	*Thyrsostachys oliveri*
10.	Short cells in single pairs rarely double	*Bambusa polymorpha, Cephalostachyum pergracile, Dendrocalamus hamiltonii, D.longispathus, Melocanna bambusoides,*

		Oxytenanthera abyssinica. O.nigrociliata, Thyrsostachys oliveri, Ochlandra travancorica, Pseudostachyum polymorphum, Teinostachyum dullooa.
11.	Short cells in single and double or triple pairs.	*Bambusa arundinacea, B.nutans, B.tulda, Dendrocalamus strictus, Dinochloa maclelandii*
12.	Stomata completely over - arched by papillae.	*Bambusa polymorpha, B.tulda, Dendrocalamus hamiltonii, D.longispathus, Melocanna bambusoides. Oxytenanthera abyssinica, Ochlandra travancorica. Dinochloa maclelandii*
13.	Stomata surrounded by papillae.	*Bambusa nutans, B.arundinacea Dendrocalamus strictus Oxytenanthera nigrociliata,*
14.	Stomata with guard cells only.	*Thyrsostachys oliveri, Gigantochloa macrostachya, Teinostachyum dullooa.*
15.	Stomata with guard cells and subsidiary cells.	*Bambusa arundinacea, Dendrocalamus strictus. Oxytenanthera nigrociliata.*
16.	Short cell pair replaced by a bladder like cell here and there.	*Dendrocalamus hamiltonii.*
17.	Short cell pair replaced by unicellular microhairs in the form of spicules here and there.	*Bambusa polymorpha, Dendrocalamus longispathus, Oxytenanthera abyssinica. Thyrsostachys oliveri.*
18.	Short cell pair replaced by bicellular cylindrical microhair here and there	*Dendrocalamus hamiltonii. Melocanna bambusoides.*
19.	Short cell pair replaced by fan shaped bicellular hair here and there.	*Bambusa tulda.*
20.	Cortex very narrow.	*Bambusa arundinacea, Dendrocalamus strictus, Melocanna bambusoides, Dinochloa maclelandii*
21.	Cortex relatively wide.	*Bambsusa nutans, B.polymorpha, B.tulda, Oxytenanthera abyssinica,*

22.	Cortex homogeneous and composed entirely of thin walled parenchyma.	*Bambusa arundinacea, B.nutans. Cephalostachyum pergrocile, Dendrocalamus strictus. Oxytenanthera abyssinica. O.nigrociliata, Dinochloa maclelandii.*
23.	Cortex homogeneous and composed of thick walled parenchyma.	*Bambusa polymorpha, Dendrocalamus longispathus, Melocanna bambusoides.*
24.	Cortex distinctly heterogeneous, composed of subepidermal layer of cells forming hypodermis.	*Bambusa tulda, Dendrocalamus hamiltonii, Thyrsostachys oliveri. Gigantochloa macrostachya.*
25.	Peripheral vascular zone with scattered patches of fibres not associated with vascular bundles.	*Melocanna bambusoides.*
26.	Peripheral vascular zone with scattered patches of tracheids.	*Oxytenanthera abyssinica.*
27.	Peripheral vascular zone with reduced vascular bundles composed of a single occasionally two, metaxylem vessels completely surrounded by fibre sheath.	*Bambusa nutans, B.polymorpha, B.tulda, Oxytenanthera nigrociliata, Cephalostachyum pergracile.*
28.	Peripheral vascular zone with complete vascular bundles consisting of phloem and three xylem vessels surrounded by two separate sheaths viz phloem fibre sheath and xylem fibre sheath	*Bambusa arundinacea, B.nutans, B.polymorpha, Dendrocalamus hamilotonii, D.longispathus, D.strictus, Melocanna bambusoides, Oxytenanthera nigrociliata, O.abyssinica, Cephalostachyum pergracile, Thyrsostachys oliveri.*
29.	Peripheral zone bundles with phloem fibre sheath much smaller than the xylem fibre sheath	*Bambusa nutans, B.polymorpha, Dendrocalamus longispathus, D.strictus, Cephalostachyum pergracile, Melocanna bambusoides, Oxytenanthera abyssinica, O.nigrociliata, Thyrsostachys oliveri.*
30.	Peripheral zone bundles with phloem and xylem sheaths well developed but unequal in size.	*Bambusa arundinacea, Dendrocalamus hamiltonii, B.polymorpha.*
31.	Transitional zone bundles with phloem fibre sheath	Bambusa arundinacea, B.nutans, B.polymorpha, B.tulda, Cephalostachyum

	and xylem fibre sheath well developed but unequal in size.	pergracile, Dendrocalamus longispathus, D.strictus, Melocanna bambusoides, Oxytenanthera abyssinica, O.nigrociliata, Thyrsostachys oliveri.
32.	Transitional zone vascular bundles with phloem fibre cap present and separated from phloem fibre sheath. Xylem fibre cap present and separated from xylem fibre sheath.	Dendrocalamus hamiltonii.
33.	Transitional zone vascular bundles with phloem fibre cap present and attached to phloem fibre sheath. Xylem fibre cap present and attached to xylem fibre sheath.	Bambusa polymorpha.
34.	Transitional zone vascular bundles with phloem fibre cap present and separated from phloem fibre sheath. Xylem fibre cap present and attached to Xylem fibre sheath.	Bambusa arundinacea, Thyrsotachys oliveri.
35.	Transitional zone vascular bundles with phloem fibre cap absent but xylem fibre cap present and attached to xylem fibre sheath.	Oxytenanthera nigrociliata, O.abyssinica.
36.	Transitional zone vascular bundles with phloem fibre cap absent but xylem fibre cap present and separated from xylem fibre sheath.	Dendrocalamus hamiltonii.
37.	Transitional zone vascular bundles with phloem fibre cap absent and xylem fibre cap absent, The vascular bundles consisting of only phloem fibre sheath and xylem fibre sheath.	Bambusa tulda, Dendrocalamus *longispathus, D.strictus, Melocanna bambusoides, Oxytenanthera abyssinica, O.nigrociliata, Dinochloa maclelandii*

38.	Central and inner zone vascular bundles with only one fibre cap viz xylem fibre cap attached to xylem fibre sheath with a narrow stem like portion.	*Melocanna bambusoides, Oxytenanthera abyssinica.*
39.	Central and inner zone vascular bundles with only one fibre cap viz xylem fibre cap distinctly separated from xylem fibre sheath.	*Dendrocalamus strictus, Bambusa nutans, Dinochloa maclelandii*
40.	Central and inner zone vascular bundles with two fibre caps. Phloem fibre cap attached to the phloem fibre sheath, and xylem fibre cap separated from xylem fibre sheath.	*Bambusa polymorpha.*
41.	Central and inner zone vascular bundles with two fibre caps. Phloem fibre cap separated from the phloem fibre sheath. Xylem fibre cap sometimes attached to the xylem fibre sheath, but mostly free.	*Bambusa arundinacea, Dendrocalamus longispathus.*
42.	Central and inner zone vascular bundles with two caps both free from the sheaths.	*Bambusa arundinacea, B.tulda, Dendrocalamus longispathus, D.hamiltonii, Oxytenanthera nigrociliata, Thyrsostachys oliveri.*
43.	Central and inner zone vascular bundles with four fibrecaps all four attached to their respective sheaths.	*Cephalostachyum pergracile, Melocanna bambuoides*
44.	Inner zone vascular bundles with only sheaths present and both caps absent.	*Bambusa nutans, Dentrocalamus longispathus, D.strictus, Oxytenanthera abyssinica, Dinochloa maclelandii*
45.	Peripheral and transitional zone type of vascular bundles scattered among central and	*Bambusa arundinacea. B.nutans, B.polymorpha, Dendrocalamus hamiltonii, D.longispathus,*

	inner zone bundles,	*D.strictus, Melocanna bambusoides.*
46.	Independent patches of fibres scattered among the vascular bundles of central and inner zone.	*Dendrocalamus longispathus, Melocanna bambusoides.*
47.	Central cavity lined with selerieds.	*Cephalostachyum pergracile, Dendrocalamus longispathus.*

GROSSER and LIESE 1971

Grosser and Liese 1971 worked on 52 species of bamboos belonging to 14 genera obtained from a number of different countries in Asia. A detail study of the internodal structure of the culm was made from 1200 internodes of 250 culms. Their study centered around the fibro-vascular bundles and parenchyma of the internodes. They found that the shape of the vascular bundles in spite of variations, differed from species to species confirming the earlier finding of Pattanath and Rao (1969). In general they considered the variations to be a modification of the basic type of vascular bundle where the conducting elements are arranged in a cross supported by 'sclerenchymatous sheaths'. In this point they differ from Pattanath and Rao who described the supporting tissue of the conducting elements to be made up of long narrow thick walled fibres forming sheaths. Further Grosser and Liese recognized four basic types of vascular bundles which is reproduced here as Table. 2. Pattanath and Rao on the other hand recognized only one basic type of vascular bundle viz. the four conducting elements arranged in a cross and each supported by a crescent shaped sheath of narrow long thick walled fibres. In addition to the sheaths, there are the fibres caps ranging from one to four in number but generally two and made up of broader thinner walled fibres. All other types were considered by them as modifications of this basic type.

Table - 2 Grosser & Liese's Basic Vascular Bundle Types in Bamboo.
(Source: Grosse & Liese 1971)

Vascular bundle Type	Characteristic	Occurrence
Type - I 	Consisting of one part central vascular strand supporting tissue only as sclerenchyma sheaths. Intercellular space with tyloses	In all species with leptomorph rhizomes throughout the culm as only type (*Arundinaria Phyllostachys*)
Type - II 	Consisting of one part (Central vascular strand, supporting tissue only as sclerenchymatous sheaths.	In species with pachymorp rhizomes growing either in single culm formation (*Melocanna*) or in clumps

	Sheath at the intercellular space (protoxylem) strikingly larger than the other ones. Inter - cellular space without tyloses.	(*Cephalostachyum, Schizos tachyum, Teinostachyum*) In *Cephalostachyum* as only one type throughout the culm. In *Melocanna, Schizostachyum, Teinostachyum* in base internodes often together with type III.
Type - III 	Consisting of two parts (Central vascular strand and one fibre strand) fibre strand inside the central strand. Sheath at the intercellular space (protoxylem) generally smaller than the other ones.	In clump forming species with pachymorph rhizomes (*Bambusa, Dendrocalamus, Gigantochloa, Thyrsostachys* at the base internodes combined mostly with type IV, in the middle and upper parts as only type. In *Melocanna, Schizostachyum, Teinostachyum* combined at the base internodes with type II. In some *Oxytenanthera* species only type throughout the culm.
Type - IV 	Consisting of three parts (central vascular strand and two fibre strands); fibre strands outside and inside the central strand.	In clump forming species with pachymorph rhizomes (*Bambusa, Dendrocalamus, Gigantochloa, Thyrsostachys* mostly at the base internodes, seldom at the middle part; always combined with type III.

Grosser and Liese found that in a species two types may exist according to the position in the culm and the wall thickness. Basic types II, III and IV occur only in the culms of pachymorph genera. Basic type II exists either alone or in combination with type III, in which case type III bundles are restricted to the lower part of the culm; and type II occurs in the middle and upper parts of the culm. Type III may also occur in the middle and upper parts. Only in a very few species is found type IV in the middle and upper parts of the culm eg. *Dendrocalamus merrillianus*. They suggested a classification based on the above mentioned four basic types of vascular bundles as indicated below:-

Group A. Genera having Type II above
eg. *Arundinaria, Phyllostachys, Tetragonocalamus.*

Group B. All Genera having type II

B$_1$. Genera having only type II
eg. *Cephalostachyum.*

B$_2$. Genera having type II & type III
eg. *Melocanna, Schizostachyum, Teinostachyum.*

Group C. Genera having Type III above
eg. *Oxytenanthera.*

Group D. Genera having type III & IV
eg. *Bambusa, Dendrocalamus, Gigantochloa,Thyrsostachys.*

They divided bamboos into two groups viz. leptomorph genera, that is genera having monopodial type of rhizome, and pachymorph genera, that its genera having sympodial type of rhizome. They found that all leptomorph genera had vascular bundles of type I and belonged to group A. The pachymorph genera on the other hand comprised of group B, C and D with vascular bundles of type II, III and IV respectively. Further, they also found that in general the number and density of vascular bundles are characteristic of a species.

The main difference between the classification suggested by Grosser and Liese (1971) and Pattanath and Rao 1969 is that Grosser and Liese concentrated mainly on the vascular bundles of the central zone of individual internodes for the development of the key, although in the text of their research paper they have indicated other characteristics such as vascular bundles of peripheral and transitional zone which could be used for differentiating one species from another. Pattanath and Rao on the other hand considered a range of anatomical characters starting from the epidermis to the cortex, and the various types of vascular bundles found within any given internode from the periphery to the central cavity, and their modifications in the upper parts of the culm, and came to the conclusion that circumscription into genera was not possible from the limited number of species studied by them. Instead of a key they listed a number of diagnostic characters which could be used in different combinations for the identification of species.

BAHADUR 1979

In 1979 Bahadur working in the Forest Research Institute Dehradun India, developed a key for the identification of bamboos based on culm buds and bud sheaths as given in Table 3. Bahadur found that a series of culm sheaths that clothe successive internodes show a progressive change acropetally in size, shape, substance and vestiture of both the sheath proper and its appendages. He feels that only culm sheaths in the middle of the culm are diagnostic of a given species, as features which are typical of a species are somewhat obscured from the bottom and top of the culms. The culm buds and juvenile shoots of bamboos on the other hand are highly remarkable, and those sheaths borne nearer the rhizome as root portion vary least topographically, and sheaths borne by the young growing vegetative buds of different species has a distinct appearance.

Table-3 Characteristics of Culm Sheaths and young Shoots of some bamboo species
(Source: Vermah and Bahadur 1980).

Species	Culm Sheath Characteristics	Diagnostic Characters of Young Shoots.
Bambusa arundinacea Chromosome nos (72) (70)	Elongated, glabrous, thick, small, narrow, culms erect, blade broadly triangular branches thorny.	Metallic, purplish green. growing apex blunt.
Bambusa burmanica.	Fairly large felted asymmetric or oblique; atleast one auricle situated laterally. Culm nodes with one white band on both sides.	Yellow stripes on green surface. calcareous bands on both sides of the culm node.
Bambusa copelandii	Very large, glabrous, not conical, ligule serrate, blade narrow elongated tapering	Blade and edge of sheath very sharp, copper coloured, auricles more or less absent.
Bambusa glaucescens (*B.multiplex*) (*B.nana*) chr no.72	Conical, broad based blade smaller than sheath.	Auricles inconspicuous blade broad at the base, suddenly tapering upwards and slender.
Bambusa longispiculata	Fairly large, felted, asymmeetric or oblique. Auricles at the top. Blade triangular, accuminate, cordate at the base, auricles long ciliate.	Greyish green with dark brown hairs, blades leathery and mucronate.
Bambusa nutans	Fairly large, felted, asymmetric or oblique. Auricles at the top. Blade triangular acute, base continuous with long ciliate band and auricles.	Caterpillar like auricles. Body dark brown, apex green.
Bambusa pallida	Conical, broad-based, blade longer than the sheath.	Green with brown blades, sheaths almost sticking.

Species	Culm Sheath Characteristics	Diagnostic Characters of Young Shoots.
Bambusa polymorpha chr. no.72	Very large glabrous much broader than long, auricles present falcate.	Auricles biserrate lower blade brown other blades green cup shaped.
Bambusa tulda	Fairly large, felted, asymmetric or oblique auricles, atleast one of them situated laterally; culm nodes with a white calcareous band on one side.	Yellow stripes on green surface; calcareous band on one side of culm node.
Bambusa ventricosa	—	Dark brown, apex green, auricles distinct; internodes pitcher shaped.
Bambusa vulgaris chr. no. 72	Fairly large felted symmetric broad covering a major part of the internode. Yellow or green rounded at the top, blade triangular	Dark brown, apex green, auricles distinct. Internodes cylindric (not pitcher shaped) pure yellow culms with all intermediary shades. Yellow with green stripes most common.
Cephalosta--chyum pergracile	Fairly large, felted symmetric broad much short than the internode; chestnut brown in colour.	Chestnut brown body; blades small cup shaped, auricles with long hairs.
Dendrocalamus brandisi (chr.no.72 +2B)	Very large, glabrous, not conical, ligule entire, auricles long not plaited.	Grey with blackish effect. Blades very dark brown.
Dendrocalamus calostachys	Very large glabrous not conical, ligule entire, auricles short, plaited.	Purplish with a tinge of orange and green (pinkish). Blades dark green.
Dendrocalamus giganteus	Very large, glabrous not conical ligule serrate, blades broadly	Column very big glaucous - green

Species	Culm Sheath Characteristics	Diagnostic Characters of Young Shoots.
(chr. no. 72)	triangular.	glabrous, auricles absent
Dendrocalamus hamiltonii	Very large, glabrous, conical broad, auricles absent.	Tomentum perfectly black, auricles present, blade stiff and pointed.
Dendrocalamus longispathus (chr.no.72)	Elongated, glabrous papyraceous, very long.	—
Dendrocalamus membranaceous	Dome shaped, felted blade narrow, elongated tapering.	Sheath light brown mottled blade linear.
Dendrocalamus strictus (chr.no.72)	Dome shaped felted, blade triangular, pointed.	Brown with very thick dark brown hairs, apex short, auricles absent culms glaucous.
Dinochloa maclelandii	Elongated, glabrous, thick small, narrow. Blade lanceolate acuminate. Culms scrambling.	Blades reflexed climbing bamboo with extremely large leaves.
Gigantochloa atter	—	A very fine brown stripe along the margin of the sheath. Culms green or black.
Melocanna baccifera (*M. bambusoides*).	Fairly large felted symmetric, narrow, auricles prominent, blade arising from a concave depression.	Ligule horse shoe shaped: blades flagellate, stems single (non-clump forming).
Oxytenanthera albociliata	Fairly large felted symmetric, broad, covering a major part of the internodes, yellow or green truncate at the top, blade lanceolate and candle shaped.	Variegated, blades reflexed apex short and pointed.
Oxytenanthera abyssinica.	—	Blue-green with cream yellow blades apex shortly pointed.

Species	Culm Sheath Characteristics	Diagnostic Characters of Young Shoots.
Oxytenanthere nigrociliata.	Fairly large, felted, symmetric, broad covering a major part of the internode, yellow or green, blade conical, ligule narrow.	Shining black hairs, apex long pointed column conical.
Thyrsostachys oliveri	Fairly large felted symmetric narrow, auricles absent, blade arising from a flat top.	Yellow lines on the sheaths, blades linear. Lower culm nodes often bearded.

Vermah and Bahdur believe that though it is possible to identify individual species by the key presented in Table 3, it is impossible to determine the generic affinities, which is similar to the finding of Pattanath and Rao 1969 in their studies on the anatomical structure of eighteen species of bamboos belonging to eleven genera.

HIROSHI USUI 1985

In 1985 Hiroshi Usui working on Japanese bamboos belonging to 23 species of six genera suggested a classification based on prophylls which are responsible for regulating the branching in different species of bamboos. Earlier in 1957 Usui had published a paper titled - "Morphological studies on the prophyll of Japanese Bamboos" in the Botanical Magazine of Tokyo. He fixed buds of new shoots in Bouins Fluid and embedded the samples in paraffin wax and cut 10mm thick sections and stained them with tannic acid and ferric chloride and Haidenhains iron haematoxylin. He found that he could distinguish six types of prophylls as given below and each type was characteristic of the genus.

1. *Sasa type* : The prophyll is two keeled and the species produces only one branch.

2. *Arundinaria type:* has atleast three, two keeled prophylls and produces atleast three branches. There are species that have seven to nine branches at a node but these develop from secondary buds.

3. *Shibatea type* : The two outermost prophylls are not of the usual two keeled form. Two are separated while the others are two keeled. Shibatea produces five branches.

4. *Phyllostachys type* : The cross section of this type is the same as the right half of the shibatea type. This type usually produces two branches often three.

5. *Chimonobambusa type :* Produces three branches. These are characterised by three separate prophylls and the three branches are independent upto the base.

6. *Bambusa type* : Produces numerous branches at a node. The outermost prophyll shows the usual two-keeled form but others do not. There are several leaf groups in a bud, and their prophylls have pointed tips.

WEN TAIHUI and CHOU WENWEI 1985

In 1985 Wen Taihui and Chou Wenwei studied the anatomy of vascular bundles of 99 species belonging to 28 genera of bamboos collected form Zhejiang, Fujian, Guangdong, Yuennan, Jiangxi, and Shichuan provinces of China. They categorized vascular bundles of bamboos into five types viz double broken type, broken type, slender waist type, semi open type and open type. They found that though the vascular bundles varied much within a culm they remained relatively stable within given internodes of a species. This observation is in agreement with the observations of Pattanath and Rao (1969) who state that though the structure of vascular bundles varies much within a culm, the pattern of variation is characteristic of a species. It is also in agreement with the observations of Grosser and Liese (1971) who mention that form, size, and pattern of vascular bundles can vary so markedly that an internode from the culm base exhibits a completely different structure from one taken in the middle which in turn can differ from one taken at the top. The extent of variation in the form differs from species to species and is mainly determined by the basic types.

However unlike Pattanath and Rao 1969 who found it impossible to circumscribe generic characters, both Grosser and Liese (1971) and Wen Taihui and Chou Wenwai (1985) were able to produce generic keys. The key developed by the latter are reproduced here as Table 4.

Table 4 Wen Taihui and Chou Wenwei's generic key to the vascular bundles of bamboos

(Source : Wen Taihui & Chou Wenwei 1985)

The vascular bundles in the lower part of the culm are mostly double broken type and middle and upper part with broken type vascular bundles. Broken type and double broken type coexist. The lower part of the culm with a considerable number of vascular bundles, the upper and middle with sudden reduction of vascular bundles, the lower part with vascular bundles over twice as many as those in the upper and middle parts. ....***Thyrsostachys***

The lower part of the culm with many vascular bundles by gradually reducing to the middle part upwards. ....***Gigantochloa***

The lower part of the culm with few double broken type vascular bundles. The lower part with outer vascular bundle sheath developed particularly well and its section area is larger than the sum total of other vascular bundle sheaths......***Dendrocalamus***

The lower part of the culm with outer bundle sheath particularly well developed and its section area is as large as that of the left and right side vascular bundle sheath or similar. ....***Oxytenanthera***

The lower and middle parts or only middle parts of the culm with the double broken type vascular bundles. The culm with the broken type and without the double broken type vascular bundles. In the middle of the cross section surface of the upper, middle and lower parts of the culm, or near the inner wall, scalarifrom vessels are specially big. ....*Lingnania*

Scalariform vessels of the vascular bundles of the culm are of the common size one open type vascular bundle near the inner wall. ....*Neosinocalamus*

Two open type vascular bundles near the inner wall are changeable (a few species with double broken type vascular bundles.) ....*Bambusa*

Vascular bundles of the culm do not break, fibre strands do not proliferate at the upper and lower parts of the central vascular bundle. Slender waist type vascular bundles. The lower part of the culm with a row of over ten vascular bundles, the upper and middle parts with a sharp reduction. ....*Cephalostachyum*

The lower part of the culm with relatively fewer vascular bundles, the vascular bundles in the upper, middle and lower parts are almost equal in number. Culms slender but with fairly big vascular bundles, the inner sheaths of the vascular bundles near the inner wall are normal. ....*Melocanna*

Culms thick but with comparatively small vascular bundles, the inner sheaths of the vascular bundles near the inner wall are undeveloped. ....*Schizostachyum*

The open type and semi open type vascular bundles, with the open type as the regular vascular bundles. The undifferentiated and semi-differentiated vascular bundles are especially well developed, the sectional area of the sclerenchyma is one to three times the size of the regular vascular bundles. ....*Ferrocalamus*

The undifferentiated and semi-differentiated vascular bundles are a little bigger than regular vascular bundles. The outer vascular bundle sheaths of the semi-differentiated vascular bundle is especially well developed and is one to three times as big as other vascular bundle sheaths. ....*Pseudosasa*

The outer vascular bundle sheath of the semi-differentiated vascular bundle is almost as big as or a little bigger than other vascular bundle sheaths ....*Chimonobambusa*

All the vascular bundles are semi-open. The semi-differentiated vascular bundles are two times bigger than regular bundles. ....*Gelidocalamus*

The semi-differentiated vascular bundles and regular vascular bundles are almost equal in number. A row of about three to four vascular bundles, amphipodial rhizomes. ....*Subgenus Pleioblastus*

A row of more than five vascular bundles. The lower part of the culm with a row of five to six vascular bundles. Semi-differentiated vascular bundles and the inner vascular bundle sheath is almost rectangular. ....*Yushania*

Semi-differentiated vascular bundles and the inner vascular bundle sheath is triangular. ....*Sasa*

The lower part of the culm with a row of seven to eight vascular bundles, the inner sheath of the semi-differentiated vascular bundle is triangular, the outer sheath is oval. ....***Semiarundinaria***

The open type and the semi-open type vascular bundles coexist mainly with the semi-open type vascular bundles. The vascular bundles in the lower part of the culm are all, semi-open near the inner wall of the upper and middle parts are two or three open type vascular bundles. ....***Fargesia***

Near the inner wall of the upper, middle or lower part are one to two open type vascular bundles. The inner vascular bundle sheath of the semi-open type vascular bundle in the three parts of the culm is triangular. ....***Subgenus Amarus***

The characteristics and morphology of the inner vascular bundle sheath of the semi-open type vascular bundles are irregular. ....***Clavinodum***

Mainly with open type vascular bundles. The semi-differentiated vascular bundles and the regular vascular bundles are almost equal in size. ....***Phyllostachys***

The semi-differentiated vascular bundles bigger than regular vascular bundles. The outer sheath of the semi-differentiated vascular bundle is almost as big as the outer sheath of the regular vascular bundle. ....***Indocalamus***

The outer sheath of the semi-differentiated vascular bundle is quite well developed and far bigger than the outer sheath of the regular vascular bundle.

..........***Brachystachyum***

The open type and the semi-open type vascular bundles coexist: sometimes there are parenchyma cell rings among groups of vascular bundles or the open type vascular bundles among the semi-open type vascular bundles. The vascular bundles in the upper, middle and lower parts of the culm are in the ratio of 4:5:9 shaped like an upside down pagoda. ....***Indosasa***

The vascular bundles in the upper, middle and lower parts of the culm are in the ratio of 6:8:8 almost equal in number. ....***Sinobambusa***

Different authors have used different terminologies for describing vascular bundles in bamboos. Pattanath 1965 describes a fully differentiated vascular bundle as containing two metaxylem vessels placed adjacent to each other with a protoxylem vessel placed below, and phloem group above, forming a cross, and each conducting element supported by a crescent shaped independent sheath of narrow thick walled fibres. In addition to the sheaths are two fibre patches forming caps—one above the phloem fibre sheath known as phloem fibre cap, and the other below the protoxylem fibre sheath known as xylem fibre cap. Variations of this basic type occurs when (i) one of the caps is absent; (ii) when one of the caps is joined to its sheath; (iii) when both caps are joined to their respective sheaths; (iv) when instead of two caps there are four caps all joined to their respective sheaths.

Corresponding to the basic type described by Pattanath 1965 is Grosser Liese's Type IV and Wen Taihui and Chou Wenwei's "DOUBLE BROKEN TYPE". Corresponding to variation number, i. described above is Grosser and Liese's Type III and Wen Taihui and Chou Wenwei's "BROKEN TYPE". Corresponding to variation number ii. described above is Grosser and Liese's type II and Wen Taihui and Chou Wenwei's 'SLENDER WAIST TYPE VASCULAR BUNDLE'. Corresponding to variation number iii. described above is Grosser and Liese's type I and Wen Taihui and Chou Wenweis "OPEN VASCULAR BUNDLES". In addition to these, Wen Taihui and Chou Wenwei describes yet another type known a 'SEMI OPEN VASCULAR BUNDLES' in which both fibre caps are absent and only the four fibre sheaths are present. This emphasises the need to standardize the terminology for describing fibrovascular bundles of bamboos.

ZHANG GUANG ZHU 1985

In 1985 Zhang Guang Zhu made some studies on the chromosome number of some bamboo species (Table 5). A survey of cytogenetic studies on bamboos reveals that the emphasis so far has been on determination of chromosome numbers in different species. However a survey of chromosome numbers in angiosperms shows that genera and even families may have the same chromosome numbers. In some cases many related species have different chromosome numbers. In some instances differences involve one or two chromosomes, while in others they involve multiples of a number which is found in the gametes of some of the species and which is considered to comprise of the basic chromosome complement (x) or genome of the group.

However Zhang Guang-Zhu's studies brought out a number of interesting observations (i) He found that many bamboo species have two or more chromosome numbers in their somatic cells. The chromosome number in *Bambusa pervariabilis* for instance is $2n = 64$, and 56, and occassionally 72. This phenomenon he found was quite common in bamboos with sympodial rhizomes (ii) Further, *Bambusa vario-striatus* is a natural mutant because it has a very large chromosome number viz $2n = 96$. Though *B.vario-striatus* flowers easily the percent of seed setting is low (iii) In addition to the chromosome numbers of $2n = 72$ and $2n = 48$, many bamboo species have $2n = 64$. (iv) Even though the chromosome number of *Bambusa pervariabilis*, *Dendocalamus latiflorus* and *Bambusa textilis* is different from each other, they can be easily hybridized and have a strong affinity for each other (v) Bamboos with $2n = 72$ and $2n = 64$ can be crossed with each other, (vi) Bamboos with $2n = 64$ is fertile because it is an octoploid with a basic of 8 and an even number euploid is generally fertile. (vii) chromosome numbers of bamboos adapt to variations of temperature zones. For instance clump bamboos in tropical zones mostly have $2n = 72$, monopodial bamboos in the warm temperate zone mostly have $2n = 48$, and bamboos in the southern subtropical zone mostly have $2n = 64$. The number is in the range of 72 and 48, (viii) The difference in chromosome number is affected by climatic conditions, and they decrease gradually from the tropical zone to the subtropical zone (72—64—48). If 2n is less than 48 in woody bamboos, they are found in the colder regions viz on high latitudes, or at high elevations.

When studying a group of plants such as bamboos which exhibits such a diversity, one has to concentrate on the process and routes by which that diversity has been achieved. Chromosomes provide the physical basis by which the genetical stability and continuity of individuals and populations is maintained. This is evident through the orderly sequence of chromosome behaviour during mitosis and meiosis where their exact duplication and segregation provides a firm and consistent basis for the transmission of genetical information. Nevertheless chromosomes are also the vehicles for generating the variation which is essential for evolution to proceed. A study of its variation from species to species is bound to yield some information whereby relationships and true differences can be determined. Regular opportunities for the generation of variability are provided by crossing over during meiotic prophase. But the less predictable changes in chromosome structure are also important for variation and evolution of plants. These changes result from spontaneous chromosome breakages.

Moreover the duration of the complete cell cycle has been shown to be closely correlated with nuclear DNA content and can be measured by means of Feulgen photometry. The duration of cell cycles and its constituent phases can for example be determined by feeding titrated thymidine to actively growing roots for 30 minutes, and then by means of autoradiography counting the number of labelled prophases in the samples taken at 2-hourly intervals. Differences between taxa in the duration of the cell cycle have been repeatedly demostrated (Quastler and Sherman 1959).

Moore (1976) shows that synapsis viz the precise pairing of homologus chromosomes during zygotene is valuable for determining the degree of resemblance between chromosomes of plants brought together in hybrid combinations. It is an essential precursor for production of chiasmata formed between non situ chromatids during pachytene. The distribution of chiasmata does not occur at random, since the presence of one chiasma restricts the proximity of subsequent chiasmata, and the presence of heterochromatin restrict chiasma formation. There is abundant evidence for the genetic control of the frequency and distribution of chiasmata.

Further, the way homologous chromosomes or daughter chromosomes move in a regular manner to opposite poles during cell division to provide the physical basis of mendelian segregation is known to be clearly under strict genotypic control. For instance there is evidence that in long chromosomes the telomere acts as a fulcrum to give an easier and more regular separation. In organisms with multiple or diffuse centromeres the spindle fibres attach themselves at points along the length of the chromosomes, so that all parts move simultaneously polewards as rods at right angles to the spindle axis. The telomeres have a very similar structure to centromeres and sometimes show nonhomologous association with them. In some plants they can behave as centromeres becoming attached to, and moving on the spindle during cell division. Such neocentromeres develop only in certain genotypes, but are known in a variety of plants both at mitosis and meiosis, although the activity is much more pronounced in meiosis.

Table 5 Chromosome Numbers of Some Bamboo Species

(Source: Zhang Guang-Zhu 1985)

Name of Species	Chromosome Number
Arundinaria invatekensis	48
A. gigantea	48
A. simonii	48

Species	Chromosome number
Bambusa arundinacea	72
B. floribunda	72
B. multiplex	72
B. pervariabilis	64, 56, 72
B. lapida	64, 52
B. chungii	72, 64
(Lingnania chungii)	
B. emeiensis	70
(Sinocalamus affinis	
B. biciatricatus	64, 72
(Sinocalamus biciatrictus)	
B. dissemulater	64
(var albonodia)	
B. polymorpha	72
B. rutila	64
B. sinospinosa	64
B. tulda	72
B. textilis	64, 56, 72
B. variostriatus	96, 84
Bambusa sp. Guabgxi	64
Cephalostachyum pergracile	72
Cheng Ma Qing No.1	68
Qing Ma Cheng No.14	68
Chimonobambusa marmorea	48
C. falcata	48
Dendrocalamus brandisii	72 + 2B
D. giganteus	72
D. hamiltonii	72
D. longispathus	72
D. minor	72
D. latiflorus	72, 64, 48
D. strictus	72, 70
Gigantochloa macrostachya	72
Guadua capitaya	46
G. chacoensis	46
G. paraguayana	46
Indocalamus wightiana	48
Melocanna baccifera	72
Phyllostachys aurea	48
Ph. bambusoides	48
Ph. striata	48
Ph. marliacea	48, 72
Ph. flexuosa	54
Pleioblastus fortunei	48
P. gramineus	48
P. hindsii	48

P. communis	48
P. chino	48
P. simonii	48
P. pygmaeus	54
Pseudosasa japonica	48
Sasa kazsa	48
S. kurilensis	48
S. paniculata	48
Sasa sp (3x)	36
Sasamorpha purpurascens	48
Semiarundinaria yashadake	48
Semiarundinaria patlingii	48
Sinobambusa tooksik	48
Sinocalamus stenoauritus	68
Tetragonocalamus agulatus	48
Thamnocalamus aristatus	48

ZHANG GUANG ZHU and CHEN FU QUI 1985

In 1985 Zhang Guang-Zhu and Chen Fu Qui reported the results of their studies on bamboo hybridization. Their studies have a bearing on taxonomy because they found (i) that the hybridizing affinity of the different bamboo species under the same genus is closer; (ii) under different genera but with similar ecological characteristics varies tremendously; (iii) The hybridizing affinity of the species in different genera and with various ecological characteristics is less successful; and (iv) The percentage of seed formation in the less successful hybridizations can be raised by mixed pollination of bamboos of different genera and with various ecological characteristics.

They carried out hybridization trials of 21 groups including 4 genera and 7 species. When they crossed *Dendrocalamus minor* with *Dendrocalamus latiflorus* they obtained 22 percent seeds. *Bambusa textilis* with *Bambusa pervariabilis* produced 13.6 percent seeds. *Bambusa textilis* with *Bambusa sinospinosa* produced 10 percent seeds. On the other hand species of different genera were crossed together eg. *Bambusa pervariabilis* crossed with *Dendrocalamus latiflorus or Bambusa textilis* with *Dendrocalamus latiflorus* produced 8.1 to 14.5 percent seeds. But when *Bambusa sinospinosa* was crossed with *Dendrocalamus latiflorus* only 0.6 to 1.6 percent seeds were produced. Further when *Bambusa pervariabilis* was crossed with *Phyllostachys pubescens* 1.3 to 3.8 percent seeds were produced; *Bambusa textilis* X *Phyllostachys pubescens* yielded 1.0 to 2.0 percent seeds, *Bambusa sinospinosa* X *Phyllostachys pubescens* yielded 0.47 to 1.56 percent seeds. Further crosses were made with mixed pollen from different genera of bamboos. For instance when *B.pervariabilis* was crossed with *P.pubescens* + *D.latiflorus* 8 percent seeds were produced.

They also found that a strong hybridizing affinity indicates close relationships, and a weak hybridizing affinity indicates, that the relationship is more distant. Further they found, that *Bambusa pervariabilis* had a close affinity with *Dendrocalamus latiflorus.*

LALITA KUMARI et al 1985

These Researchers suggested a new approach to the identification of a number of different species of bamboos based on the electrophoretic resolution of peroxide isozymes. Diagnostic characters of these species are presented in the form of a key and *Bambusa nutans, B. glaucescens, B. polymorpha, B. arundinacea, B. longispiculata, B. tulda* are grouped together and they are separated from *B. vulgaris, B. ventricosa, B. pallida* and *B. balcooa*.

THE CURRENT DECADE (1986-1996)

In this decade there has been a tremendous increase in the utility of bamboos throughout the world for various small scale as well as large scale industries. This has triggered off an interest in bamboo identification and classification and establishment of live collections of bamboos termed as bambusetums. Moreover several bamboo societies have also been founded in different parts of the world including Japanese Bamboo Society, American Bamboo Society, European Bamboo Society, Indian Bamboo Society etc. and Bamboo Information Centres have been created in China, India etc. This has resulted in a large number of publications, some dealing with descriptions of new species (Naithani 1992, 1993), nomenclature of some species of bamboos (Naithani 1990, 1991, 1994) revision of genus and species of bamboos (Naithani 1990, Naithani and Bennet 1991) and taxonomic studies (Bennet and Gaur 1990) etc.

Stapleton (1989) studied Bhutanese bamboos and attempted the preparation of a key for the identification of Bhutanese bamboos using vegetative characters and grouped them into 15 genera.

Bennet and Gaur (1990) worked on the taxonomy of 37 important bamboo species and used vegetative characters such as morphology of young shoots, terminal buds, branching patterns and nodal region for identification of the different species taken up for the study.

Naithani and Bennet (1991) made an attempt to revise the placement of different bamboo species. They tried a merger of seven Burmese bamboos viz. *Schizostachyum burmanicum, S. distans, S. scandens, S. strictum, S. tavoyanum, S. virgatum* and *Dendrocalamus detinens* with *Cephalostachyum, Dendrochloa* and *Neohouzeaue*.

Some studies on the anatomical structure of bamboo leaves are also being carried out in the Forest Research Institute, Dehradun. These studies however are at a very preliminary stage and nothing concrete appears to have come out of them as they consist merely of some empirical data.

DISCUSSION ON DIFFERENT CLASSIFICATION SYSTEMS

A review of the classifications of bamboos presented so far, reveal that different researchers have used different methods, and each has a number of good and bad points, and there remains an element of personal choice in the classification which is adopted. For general purpose classifications to be really good, they have to be predictive. For instance the mention of family Poaceae at once brings into mind the peculiar floral morphology, leaf and stem anatomy, and unique seed structure, of its members, and these characteristics are likely to be present even in respect of a member of Poaceae which is not yet investigated. Thus the statement that a plant belongs to Poaceae, if correct, automatically predicts a number of basic features that will be applicable to the plant. A classification which enables such an inference is a predictive classification, and predictivity is one of the most obvious criteria that can be used to judge a classification system. Classifications which have high levels of predictivity are termed as natural classifications, and the groups which they delimit are natural groups.

There are also many sorts of special purpose classifications that are very low in predictivity, and which comes under the category of artificial classification. An example of this among the different systems described is the classification based solely on prophylls, solely on chromosome numbers, or solely on fibro-vascular bundles. The main aim of such a classification is to fix the identity of a species a when only a limited sample consisting of only a part of a plant is available. Although for general purpose, predictivity is the criteria, there are other special purposes that has to have other criteria.

Since modern classifications are ideally based on a very wide range of characters, incorporation of which is a gradual and continuous process, it follows that the naturalness and predictivity of a classification depends on the extent to which a group has been investigated. The level of knowledge on bamboo species varies a great deal because in some species detail anatomical investigations have been carried out, while in others some cytogenetical characters have been investigated in addition to the existing morphological descriptions.

Thus for instance in all the early classifications based on morphological structures, *Bambusa* and *Dendrocalamus* have been placed under two different subtribes by Gamble 1881, Bentham 1883, Stapf 1897 and Hooker 1897. But Holttum in 1956 found that there was a great similarity between *Bambusa* and *Dendrocalamus* especially in the structure of the ovary. Pattanath (1965) and Pattanath and Rao (1969) pointed out that in their studies *Bambusa* and *Dendrocalamus* have several anatomical features in common. In Grosser and Liese's (1971) classification of fibrovascular bundles also, *Bambusa* and *Dendrocalamus* occur together under group D. Further Zhang Guang Zhu (1985) also found that *Bambusa pervariabilis* and *Dendrocalamus latiflorus* and *Bambusa textilis* could easily be hybridized though they may vary in chromosome number.

Holttum also felt that Dinochloa which was placed along with *Schizostachyum* by Bentham (1863) Stapf (1897) and Hooker (1897), should be transferred and placed along with *Bambusa*, based on the ovary type. Pattanath (1965) observed that *Dinochloa maclelandii* was very similar to *Bambusa nutans* in a number of anatomical characters such as culm epidermis, fibrovascular structures etc.

Furthermore Holttum found that *Pseudostachyum*, *Teinostachyum* and *Cephalostachyum* of Bentham's subtribe 3 were very similar to *Schizostachyum* placed in Bentham's subtribe 4. Pattanath (1965) felt that *Bambusa polymorpha* differed in a great many anatomical characters from the rest of the *Bambusa* species and should be transferred to another group. All these are indications that as more information from different aspects of a species comes in, there will be changes in the classification of bamboos.

An ideal plan for a systematic investigation of bamboos for their classification would be as represented below:-

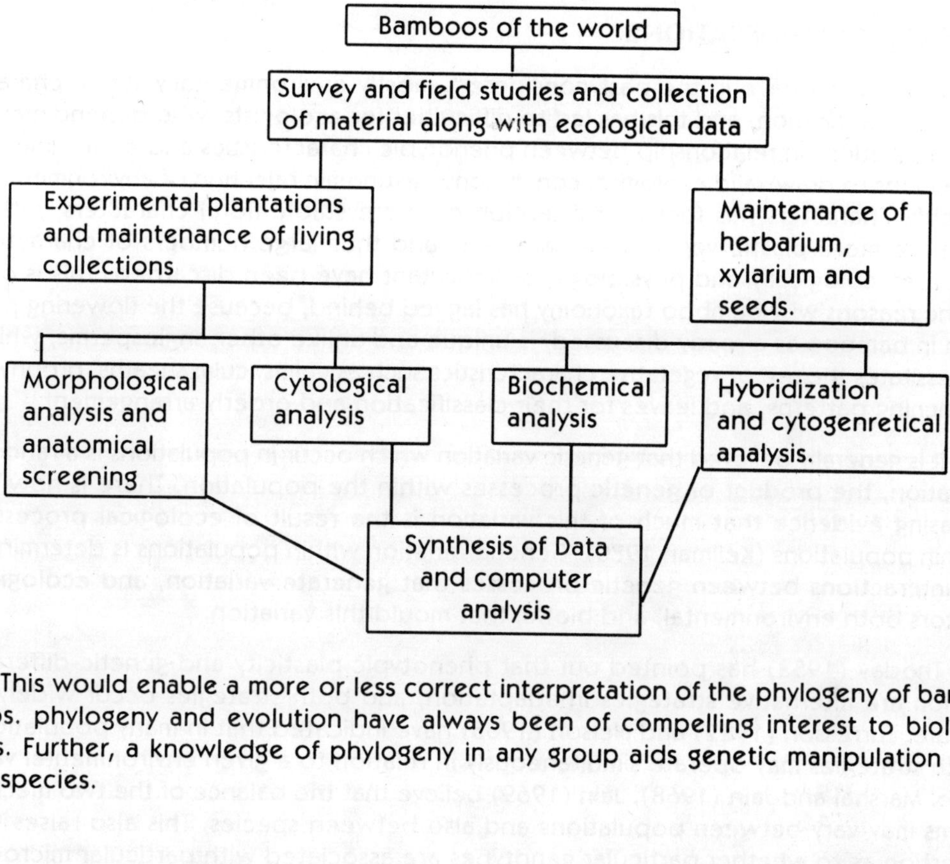

This would enable a more or less correct interpretation of the phylogeny of bamboos. phylogeny and evolution have always been of compelling interest to biologists. Further, a knowledge of phylogeny in any group aids genetic manipulation of the species.

GEOGRAPHICAL DISTRIBUTION AND ECOLOGY

GENERAL CONSIDERATIONS

Taxonomists have in the past considered, mostly environmentally stable characters for classification, and this has led to difficulties for ecologists, who depend mainly on information on relationship between phenotypic characteristics and environmental conditions to draw valid ecological conclusions. Taxonomic rejection of environmentally variable characters has focussed attention on more stable floral characters, rather than the more plastic vegetative characters, and thus large numbers of characters that are ecologically and physiologically important have been discarded. This is one of the reasons why bamboo taxonomy has lagged behind, because the flowering pattern in bamboo as already discussed, is unique and unlike other angiosperms, which necessitates the use of vegetative characteristics such as culms, culm sheaths, prophylls, branching patterns, and leaves for their classification and orderly arrangement.

It is generally believed that genetic variation which occur in populations is a random variation, the product of genetic processes within the population. There is now increasing evidence that much of this variation is the result of ecological processes within populations (Kellman 1980). Genetic variation within populations is determined by interactions between genetic processes that generate variation, and ecological factors both environmental, and biotic, that mould this variation.

Thoday (1953) has pointed out that phenotypic plasticity and genetic differentiation are alternative strategies in adaptation, and both strategies occur widely in nature. Turresson (1922) and Nelson (1965) have indicated that in many populations both strategies may operate simultaneously in relation to a given environmental variable. Marshal and Jain (1968), Jain (1969) believe that the balance of the two mechanisms may vary between populations and also between species. This also raises the question as to whether particular genotypes are associated with particular microenvironments. Some evidence on this subject has come from studies of population differentiation at the sharp boundaries between contrasting environments, and it has been observed that differences closely follow the sharp changes in environmental conditions, and are apparently adaptive (Jain and Bradshaw 1966; Antonovics and Bradshaw 1970). Genetic variation within populations has generally been studied at the morphological level. It needs however to be studied also at the physiological and biochemical level, as recent studies on allozyme and other biochemical polymorphisms have demonstrated (Fairbrother 1968, Turner 1970). This again emphasises the use of data from a wide range of characters to produce a predictable classification of bamboos.

Patterns of taxonomic diversity may be studied from a number of different viewpoints such as evolutionary, ecological, genetical etc. The interest in evolution was stimulated by Darwin in the nineteenth century, and since then, it has had an immense influence on taxonomy. It is only in the last few decades that genetics has influenced taxonomy in the form of cytotaxonomy (Moore 1968) and reproductive isolation (Solbrig 1968). The influence of ecology, physiology and biochemistry is even more recent. Ecologically important functional characters whether morphological, physiological or biochemical are now increasingly being used by numerical taxonomists.

Relationship between patterns of morphological variation in plants and ecological factors with which they are associated are most apparent below the level of the species where divergence has been more recent. Below the level of the species, variation has been studied by genecologists and biosystematists, and the emphasis has tended to be on functional ecologically important characters.

Turresson (1922) first conducted some experimental transplantations, and observed that even within morphologically homogeneous species, there may be consistent genetically fixed physiological differences, representing divergent evolutionary adaptation to the differing environments present throughout their ranges. He termed these physiological races as ecotypes. The existence of ecotype differences at both local and regional scales within a few species is now widely documented. Olmsted (1944) demonstrated that in some species there is a more gradual change in physiological properties, and this phenomenon has been termed as ecoclinal variation. Clines are presumed to reflect both gradual changes in selective pressures of the environment, and decreasing frequencies of gene interchange with increasing distances between local populations. Some ecotypic differences may be accompanied by slight morphological differences and hence may be visually detectable, but frequently uniform morphology prevails throughout the species. Data concerning environmental conditions, ecologically associated taxa, and ecological distribution are very limited and not easily available for bamboos. However some recent information on geographical distribution of bamboo species in China, Japan, Indonesia, Malaya, Thailand, India, Nepal, Srilanka and Africa are available in status reports of bamboo research of these countries.

WORLD DISTRIBUTION OF BAMBOOS

According to current literature there are 75 genera of bamboos comprising of about 1250 species excluding the ones not yet described and named. They vary in height from 15cm to over 30 metres and occur mostly in Asia and south America and to a limited extent in Africa. In mode of growth they are either the caespitose type i.e. the clump forming type such as *Bambusa*, *Dendrocalamus* and *Gaudua*, or they are the dumetose type with the single culms laterally distributed, such as *Melocanna*, *Arundinaria*, and *Phyllostachys*. They have a wide distribution in tropical and subtropical regions of the world. Their widest occurrence is in Asia followed by south America. The genera found in south America are quite distinct from those of Asia except the pan tropical genus *Arundinaria* which is found in both continents. They are an important component of the vegetation of dry deciduous, moist deciduous, and wet evergreen or tropical evergreen rain forests occurring either as an understorey, or as pure compact blocks of bamboo brakes (Qureshi and Deshmukh 1962). In Asia the largest number of species occur in China and India. They are also found in Nepal,

Sikkim, Srilanka, Burma, Pakistan, Cambodia, Thailand, Indonesia, Malaya, Laos, Vietnam, New Guinea, Philippines, Korea and Japan. The bamboos of south America have been reported to be about 179 species with some of the important ones being *Guadua angustifolia* in Columbia and Ecuador, *G. amplexifolia* in Venezuela to Nicaragua and Honduras, *G. inermis* in Mexico, *G. superba* in Brazil, and about 71 species of *Chusquea* at higher altitudes of central and south America. The species that are prominent in Asia are species of *Bambusa, Dendrocalamus, Melocanna, Thyrsostachys, Gigantochloa, Schizostachyum,* and *Phyllostachys.* According to Sharma (1985) there are about 130 species of bamboos in India, 33 species in Bangladesh, 50 species in Thailand, 12 species in Malaysia, 55 species in Philippines, 31 species in Indonesia, 26 species in Papua New Guinea, and 300 species in China.

In tropical natural forests bamboos generally form an understorey or mixture with other tree species. There are no pure natural bamboo stands except the dense *Phyllostachys* species which occur in temperate countries. Assessment of bamboos involving information on the extent of area of each bamboo species, its density and stocking, its extent and proportion, and total availability are yet to be carried out. Some efforts in this direction are being made in different countries since the last decade or so.

However a survey of the occurrence of different bamboo species is fraught with problems because of (i) the nature of gregarious flowering and subsequent deaths of bamboo clumps following seeding; (ii) bamboos occur as an understorey due to which aerial survey becomes impossible. The density therefore has to be determined only by ground surveys and sample enumeration.

ECOLOGICAL PREFERENCES OF BAMBOOS

Climatically bamboos prefer regions of high rainfall ranging from 1270 mm to about 6350 mm. But some species such as *Dendrocalamus strictus* also occur in dry deciduous forests with low rainfall (760 mm to 1000 mm) in Udaipur, Chittorgarh, and Banswara forest divisions of Rajasthan. Rainfall plays a very important role in the distribution of different species of bamboos, as well as their growth. The different species also require a set of optimum conditions for their growth and development such as temperature, humidity, soil structure, soil drainage, soil moisture, altitude and physiographic features. This is presumed to be the reason why each species occupies a characteristic habitat and rarely occurs in mixture with other species of bamboos, exceptions being *Bambusa tulda* in association with *Cephalostachyum capitatum* and *Oxytenanthera* species with some reeds (Qureshi and Deshmukh 1962).

Bamboos become luxuriant on deep loamy soils, sandy loams, and fertile clayey loams though some species such as speices of *Oxytenanthera* occupy hill tops and plateaus, with a rather depleted soil layer. Some species such as *Dendrocalamus strictus* and *Bambusa arundinacea* flourish along river banks, brooks, streams and low level depressions. The ecology and habitat of different species of bamboos with their requirements for a wide range of moisture and soil is a subject which requires some intensive study. For instance *Dendrocalamus longispathus* occurs along ravines, while *Cephalostachyum pergracile* prefer sites between dry and moist, and *Teinostachyum helferi* prefer moist humid valleys in ever green forests.

Bamboos have not only site preferences, but also altitudinal zonation as is evident in their distribution pattern in the region of Bengal, north east Himalayas and Assam collectively where in general *Dendrocalamus hamiltonii* occurs in the north, *Bambusa tulda* in the middle regions, and *Melocanna bambusoides* in the south gregariously. Further, *Bambusa tulda, B.balcooa* and *B. arundinacea* occur in the plains and valleys of lower Bengal, while *Dendrocalamus hamiltonii* occurs on lower hills and tarai belt. Then ascending upto 1220 metres are *Bambusa nutans, Dendrocalamus sikkimensis* and *Arundinaria intermedia,* and between 1200 metres and 1830 metres are *Cephalostachyum capitatum, Pseudostachyum polymorphum, Arundinaria hookeriana, Teinostachyum dullooa,* and *Dendrocalamus patellaris.* Still higher up in Darjeeling between 1830 to 2745 metres occurs *Arundinaria racemosa.* But around 2438 metres *A. racemosa* gets smaller and occurs with two other species viz. *A. aristata* and *A.falconeri.*

BAMBOOS OF CHINA

According to Shi Quantai and Chao Ching-ju (1980) there are 26 genera and more than 300 species of bamboos in China distributed on the plains, hills and the mountains at altitudes upto 3000 metres in 22 provinces. The major bamboo forest regions are south of 25° North latitude and consist of *Bambusa, Lingnania, Dendrocalamus, Sinocalamus, Schizostachyum, Dinochloa,* and *Pseudostachyum.* In these regions, the southeast coast and Taiwan province are the subregions of the southeastern monsoon forests where *Sinocalamus latiflorus, Bambusa textilis, Bambusa pervariabilis, Bambusa spinosa,* and *Sinocalamus oldhami* are the important economic species. South of Yunnan province and south east of Tibet are the subregions of the southwestern monsoon bamboo forests consisting of speices such as *Dendrocalamus strictus, Sinocalamus giganteus, Oxytenanthera fetix.* etc.

From the Yangtze River to Nanling is the largest area among all bamboo forests. Bamboos are abundant here and the primary species is *Bambusa* and *Sinocalamus* of fascicular type, *Pleioblastus* of mixed type, *Phyllostachys* of dispersive type, and a few other bamboo species. The Sichuan basin is the most typical mixed bamboo forest region, as well as the distribution centre of *Sinocalamus affinis.* West of Fujian province and in Zhejiang, Hunan and Jiangxi is found the most economically important bamboo species of China viz. Phyllostachys pubescens.

From the Yangtze River to the Yellow River at 33° - 37° North latitude is the dispersive bamboo region, which include the temperate zone of the middle and downstream sections of the yellow river, as well as the temperate zone of Shaanxi, Gansu province, and Ningxia Hui Autonomous Region, where the major bamboo species are *Phyllostachys* of the dispersive type, and *Pleioblastus* of mixed type. This region contains the major irrigated bamboo forests. On the plains, the major bamboo species are *Phyllostachys glauca, P.flexuosa, P. nuda,* and *P. bambusoides.* In the alpine areas in Shaanxi, Gansu Province, and Ningxia Hui Autonomous Region, there are large *Semiarundinaria nitida* and *Thamnocalamus spathaceus* forests.

BAMBOOS OF INDIA

In India, bamboos have a very wide range of distribution and are found as an understorey in many types of moist tropical forests such as southern hill top tropical evergreen forests, upper Assam valley tropical evergreen forests, west coast semievergreen forest, cachar semievergreen forests, cachar tropical evergreen forests, Andaman moist deciduous forests, southern mixed moist deciduous forests, and dry bamboo brakes. Similarly they occur in different types of dry tropical forests, montane subtropical forests, montane temperate forests, and alpine forests. Champion (1938) makes special reference to bamboo brakes, where bamboos are found almost as a pure crop or with a sprinkling of other timber species. These bamboo brakes are found in moist tropical forests, dry tropical forests, and montane temperate forests. Champion also mentions secondary bamboo brakes which is found in Assam. These brakes are edaphic subclimax which are ascribed to biotic factors. In moist deciduous forests bamboo brakes overlap with some of the important timber species like teak and sal and same is the case with tropical and temperate evergreen forests.

Bamboos occur in all states of India except Jammu and Kashmir, and form rich belts of vegetation in well drained parts of tropical and subtropical habitats, and grow from sealevel in the coasts, to altitudes upto 3700 metres in the Himalayas. India has a large number of indigenous species of bamboos and some have also been introduced from China and Japan. About 50 per cent of the bamboos occur in the eastern region viz. Arunachal Pradesh, Assam, Manipur, Meghalaya, Mizoram, Nagaland, Sikkim, Tripura and West Bengal. Other areas rich in bamboos are Andamans, Bastar region of Madhya Pradesh, and Western Ghats. The principal genera of bamboos occurring in India are *Arundinaria, Bambusa, Cephalostachyum, Chimonobambusa, Dendrocalamus, Dinochloa, Gigantochloa, Indocalamus, Melocanna, Neohouzeoua, Ochlandra, Oxytenanthera, Phyllostachys, Pseudostachyum, Schizostachyum, Semiarundinaria, Sinobambusa, Teinostachyum and Thamnocalamus. Besides,* two exotic genera *Pseudosasa* and *Thyrsostachys* are found in cultivation.

Arundinaria is the genus which occurs at higher elevations in the Himalayan tracts from 609 metres upto 3658 metres. *Arundinaria hookeriana* occurs from 609 to 1524 metres, *A.intermedia* from 609 to 2134 metres, *A.polystachya* from 914 to 1524 metres, *A. maling* from 1524 to 2743 metres, *A.falconeri* from 1828 to 2438 metres, *A. racemosa* from 2134 to 3658 metres, *A. griffithiana* and *A.pantlingi* from 2438 to 3048 metres, *A. aristata* from 3048 to 3658 metres. Other species of *Arundinaria* include *A.clarkei* found in Manipur and adjoining areas, *A.gracilis* in Eastern Himalayas, *A.hirsuta* in Khasi and Naga hills from 1525 to 3000 metres, *A.manii* in Jaintia hills at about 900 metres, *A.microphylla* in Sikkim and Khasi hills from 2400 to 3100 metres, *A. rolloana* in Naga hills from 1525 to 2300 metres, and *A.suberosa* in Sikkim. Khasi, and Jaintia hills from 1200 to 1500 metres.

One of the dominant genus of bamboos is *Bambusa* which has a number of speices scattered in different parts of India. *Bambusa arundinacea* occurs throughout India in the plains ascending to 1250 metres, *B.atra* also known as *B.lineata* occurs in the Andaman Island in marshy coast forests, B.auriculata occurs in Assam, *B.balcooa* in eastern Uttar Pradesh upto 600 metres, *B.khasiana* occurs in Khasi, Jaintia and Naga hills in Manipur upto 1250 metres. *B.longispiculata* occurs in Meghalaya, *B.mastersii* in Dibrugarh in Assam, *B.nutans,* in the sub Himalayan tracts from Jamuna eastwards i.e. Assam, Bengal, Sikkim and Arunachal Pradesh from 600 to 1500 metres. *B. orientalis* occurs in Orissa and Bengal, *B.pallida* in north Bengal, Sikkim, Arunachal Pradesh and Khasi hills upto 1250 metres, *B.polymorpha* occurs in Bengal and Assam, *B.schizostachys* in south Andamans, B.spinosa in the Circars and hills of south India, *B. teres* in Bengal and Assam, *B.tulda* in eastern India and northern *Circars* from 450 to 600 metres and *B.vulgaris* pan tropical commonly cultivated for ornamental purposes.

Apart from these there are a number of exotic species of Bambusa such as *B.burmanica, B.copelandi* and *B.oliveriana* introduced from Burma and cultivated in Dehradun and Calcutta, *B.glauscens* (also known as *B.multiplex* and *B.nana*) and *B.ventricosa* introduced from China and Japan and often cultivated in Assam.

There are about seven species of *Cephalostachyum* which occurs in India viz. *Cephalostachyum capitatum* occurs in Sikkim, Arunachal Pradesh and Meghalaya at altitudes of 600 to 2450 metres, *C.flavescence* occurs in Andaman Islands, *C.fuchsianum* occurs in Arunachal Pradesh and Naga hills from 1800 to 2450 metres, *C.latifolium* occurs in Sikkim, Arunachal Pradesh, Naga hills and Manipur upto 2300 metres, *C.pallidum* occurs in Arunachal Pradesh, Khasi hills and Manipur upto 1500 metres, *C.pergracile* occurs in Bihar, Assam, Naga hills, Madhya Pradesh and Arunachal Pradesh.

A number of *Chimonobambusa* species occurring in India were originally identified as *Arundinarias* such as *Chimonobambusa callosa* originally identified as *Arundinaria callosa* occurs in Arunachal Pradesh and Khasi hills from 1200 to 2280 metres, C.*densifolia* earlier identified as *A.densifolia* occurs in Anamalai hills at 2600 metres, *C.falcata* earlier known as *A.falcata* occurs in western Himalayas from Ravi to Nepal at altitudes of 1200 to 2300 metres, *C.griffithiana* earlier known as *A.griffithiana* occurs in Arunachal Pradesh, Naga, Khasi and Jaintia hills from 900 to 1372 metres. *C.hookeriana* earlier known as *A.hookeriana* occurs in Sikkim, Arunachal Pradesh, and Khasi hills from 1200 to 2450 metres, *C.intermedia* earlier known as *A.intermedia* occurs in Sikkim, and Arunachal Pradesh at 1200 to 3050 metres. *C.jaunsarensis* earlier know as A.*jaunsarensis* occurs at the source of Pindar river in Kumaon at 1800 to 3300 metres. *C. khasiana* earlier known as *A. khasiana* occurs in Sikkim, and Khasi hills at 1525 to 1830 metres. *C.polystachya* earlier known as *A.polystachya* occurs in Sikkim, and Khasi hills at 900 to 1500 metres (Vermah and Bahadur 1980).

Another important Indian genus of bamboo is *Dendrocalamus* of which there are twelve species viz. *Dendrocalamus brandisi* a native of Burma is cultivated in the Andamans, Calcutta and Dehradun, *D.calostachyus* also a native of Bruma is cultivated at Calcutta and Dehradun, D.*colleltianus* native of Burma is cultivated at Calcutta, *D. giganteus* is a native of Burma and Malaya but frequently cultivated in India *D.hamiltonii* occurs in west, central and east India in the lower hills from Simla eastward extending to upper Burma upto 900 metres. *D.hookeri* occurs in east Himalayas, Khasi, Jaintia and Naga hills from 600 to 1500 metres. *D.longispathus* occurs in Bengal and other parts of east India, *D.membranaceous* is a navtive of Burma but is cultivated in Dehradun and Calcutta, *D.patellaries* occurs in north Bengal, Sikkim, Naga hills and Arunachal Pradesh at 1200 to 1500 metres. *D.sikkimensis* occurs in Sikkim, Arunachal Pradesh, Garo and Naga hills from 1200 to 1850 metres. *Dendrocalamus strictus* is found in deciduous forests all over India except in north Bengal, Assam and moist regions of the west coast. There are a number of varieties of *Dendrocalamus strictus* such as variety *argentia*, variety *prainiana*, and variety *sericea*.

Three species of *Dinochloa* occurs in India viz. *Dinochloa andamanica* also known as *Dinochloa tjankorreh* occurs in Andaman and Nicobar Islands, *Dinochloa compactiflora* also know as *Melocalamus compactiflorus* occurs in Bengal, Assam, and Meghalaya upto 1850 metres. *Dinochloa maclelandii* occurs in Bengal and Assam.

Two species of *Gigantochloa* occur in India viz. *G.macrostachya* in Assam and Garo hills, *G.tekserah* occurs in Garo hills. In addition to these two species G.atter a native of Malaya is cultivated at Calcutta, G.verticillata a native of Malaya and Burma is also cultivated at calcutta.

Two species of the genus *Indocalamus* occur in India viz. *Indocalamus walkerianus* also known as *Arundinaria walkeriana* occur in Pulney hills in south India upto 1500 metres. *I.wightianus* also known as *Arundinaria wightiana* occurs in Nilgiris, Palghat and Tinnevelly in south India from 1800 to 2600 metres. A variety of this species viz. variety *hispidus* occurs in Nilgiri hills from 2100 to 2300 metres.

Only one species of *Melocanna* occur in India viz *Melocanna baccifera* also known as *Melocanna bambusoides* occurs in Bengal, Assam, Meghalaya, Tripura, Mizoram and other parts of east India in the plains and lower hills.

Two species of *Neohouzeoua* occur in India viz. *Neohouzeoua dulloa* also known as *Teinostachyum dulloa* occurs in North Bengal, Sikkim, Khasi and Jaintia hills. *Neohouzeoua helferi* also known as *Teinostachyum helferi* occurs in Garo hills, Khasi and Jaintia hills at altitudes of 900 to 1250 metres.

Nine species of *Ochlandra* occur in India viz. *Ochlandra beddomei* in Wynaad in Kerala, *O.ebracteata* in Parithipally Range, Kottur Reserve Trivandrum Division, Kerala, *O.scriptoria* also known as *O.rheedi* in west coast of Malabar Kerala at low elevations on river banks, *O.setigera* occur in Nilgiri hills at Godalur at about 900 metres, *O.sivagiriana* also known as *O.rheedii* var. sivagiriana Gamble, occurs in Pulney and sivagiri hills at 1200 to 2400 metres *O.talbotii* also known as *O.rheedii* var *sivagiriana* Talbot occurs in north Kanara on river banks, *O.travancorica* occurs in the plains and hills of south India and Tinevelly upto 1550 metres, *O.travancorica* variety *hirsuta* occurs in Kerala hills, *O.wightii* also known as *O.brandisii* occurs in Tinnevelly Ghats at Courtallum (Kerala hills) at low elevations upto 1100 metres.

Five species of *Oxytenanthera* occur in India and two are introduced. *Oxytenanthera bourdillonii* occur in the Ghats of Kerala at 900 to 1550 metres, *O.monadelpha* also known as *O.thwaitesii* occur in the hills of Kurnool, hills of western Ghats from Nilgiri southwards from 1050 to 1850 metres, *O.nigrociliata* occurs in Bihar, Orissa, Garo hills, Coorg, S.Kanara, Andaman and Nicobar Islands. *O.ritcheyi* also known as *O.monostigma* occurs in the west coast, and also in the western Ghats from Konkan to Anamalai hills. *O.stocksii* occurs in Konkan coasts and Ghats of north Kanara. Apart from these there are two exotics viz. *Oxytenanthera albociliata* a native of Burma but widely cultivated in Bengal, and *O.abyssinica* a native of the rocky hills of Sudan and cultivated at Dehradun.

Two species of *Phyllostachys* are native to India and three are introduced. *Phyllostachys assamica* also known as *P.bambusoides* sensu Gamble occurs in Arunachal Pradesh at 2400 metres. *P.manii* occurs in Naga hills and is cultivated at Khasi hills upto 1500 metres. Three other species of this genus are introduced into India viz. P.aurea from Japan is cultivated at Dehradun and Mussoorie. *P.bambusoides* form China or Japan has run wild in Sarahan, Upper Bashar and Himachal Pradesh at 2438 metres, *P.puberula* from Japan is cultivated in the hill stations.

One species of *Pseudostachyum polymorphum* is indigenous to India and occurs in north Bengal, Sikkim, Garo and Naga hills and Manipur upto 900 metres and another species *Pseudostachyum japonica* also known as *Arundinaria japonica* from Japan is cultivated in Darjeeling.

Simiarly one species of *Schizostachyum* viz. *S.regersii* is indigenous to India and occurs in Andamans. The other species is an exotic viz. *S.brachycladum* a native of Malaya is cultivated at Calcutta.

There are also a number of miscellaneous species of bamboos which occur in India such as (i) *Semiarundinaria patlingii* occurs in Arunachal Pradesh at altitudes of 3000 to 3350 metres, (ii) *Sinobambusa elegans* also known as *Arundinaria elegans* occurs in Naga hills from 1525 to 2300 metres, (iii) *Teinostachyum beddomei* also known as *Teinostachyum wights* occurs on the slopes of western Ghats from north Kanara to Cape Comorin at 900 to 1550 metres and in the Nilgiris. *Teinostachyum griffithii* occurs in Assam, Meghalaya and Arunachal Pradesh, (iv) *Thamnocalamus aristatus* also known as *Arundinaria aristata* occurs in central Himalayas to Arunachal Pradesh at 2700 to 3350 metres. *Thamnocalamus falconeri* also known as *A.falconeri* occurs in Jaunsar to Arunachal Pradesh at altitudes of 2250 to 2750 metres. *Thamnocalamus prainii* known also as *Arundinaria prainii* occurs in Naga and Jaintia hills from 1000 to 2240 metres. *Thamnocalamus spathiflora* occurs in western Himalayas from Sutlej through Nepal to Arunachal Pradesh at 2250 to 3050 metres. (v) *Thyrsostachys oliveri* is a native of Burma but is cultivated in Dehradun and Calcutta and *Thyrsostachys siamensis* is a native of Burma and Thailand and is cultivated in different parts of India.

In the Alpine regions of India a very limited number of bamboos occur, the main species being *Arundinaria microphylla, Arundinaria racemosa, Chimonobambusa jaunsarensis, Semiarundinaria pantlingii* and *Thamnocalamus aristatus.* The Alpine regions of the Himalayan mountains are at altitudes higher than 3050 metres. The temperate regions where bamboos occur are distributed in Himachal Pradesh, Uttar Pradesh, west Bengal and Arunachal Pradesh in the north and Tamilnadu in the south. The northern temperate region is characterised by freezing temperatures in winter months from December to February, and the temperature is determined more by altitude than latitude. Annual rainfall is 2000 mm and dense mist prevails during the monsoons. The bamboo species growing wild in this region are *Arundinaria, Chimonobambusa, Phyllostachys, Semiarundianaria, Sinobambusa* and *Thamnocalamus*. The northern temperate forests have been classified into wet, moist and dry types and occur between 1500 and 3050 metres altitude. The southern temperate forests cover the higher hills of Tamilnadu and Kerala from about 1500 metres upwards and are popularly termed as Sholas. They are evergreen forests of close canopy and are generally developed in sheltered sites on steep slopes. The climate is equable and the region gets rainfall from both the monsoons. The main bamboo species of this region are *Bambusa, Chimonobambusa, Dendrocalamus, Indocalamus* and *Oxytenanthera*.

The subtropical region is confined to the outer Himalayas and encompasses the outer reaches of Khasi, and Jaintia hills, the higher hill tops of central Indian highlands, and the annular rings circumscribing temperate areas of higher tops of south India. The altitude ranges roughly from 1000 to 2000 metres. The main genera of bamboos found here are *Arundinaria, Bambusa, Cephalostachyum, Chimonobambusa, Dendrocalamus, Dinochloa, Gigantochloa, Indocalamus, Neohouzeoua, Ochlandra, Pseudostachyum, Sinobambusa, and Thamnocalamus.*

The tropical moist region is divided into: (i) tropical wet evergreen; (ii) tropical semi evergreen; (iii) tropical moist deciduous and (iv) littoral and swamp forests. Tropical wet evergreen forests are distributed along the western Ghats in Maharashtra, Karnataka, Tamilnadu, Andamans, West Bengal, and Assam. Bamboos are usually present in the wet evergreen areas of the north east. The tropical semi evergreen forests are better developed in the north than in the south and occurs in Meghalaya, north Bengal and Orissa. Tropical moist deciduous forests cover a large part of India and occur in the Andamans, Andhra Pradesh, Assam, Arunachal Pradesh, Tripura, Bihar, Orissa, Madhya Pradesh, Maharashtra, Gujarat, Uttar Pradesh, West Bengal, Karnataka and Kerala. The temperature in some of these areas is modified by proximity to the sea and though there is general similarity in climate and vegetation, rainfall and locality factors vary widely and these areas are well represented by a wide variety of bamboo species. For instance in the tidal swamp forest of Andamans is found *Bambusa atra (B.lineata)*. The genera *Bambusa, Dendrocalamus, Cephalostachyum, Melocanna, Ochlandra, Oxytenanthera, Schizostachyum* and *Teinostachyum* are found in the tropical moist region.

In the dry tropical region which extends from the edge of the Himalayan foot hills to Cape Comorin and is bounded by the Rajasthan desert on the north west, and western Ghats on the south west are found two main species of bamboos viz. *Bambusa arundinacea* and *Dendrocalamus strictus*. Another common species is *Cephalostachyum pergracile*. The rainfall in this area varies from 250 to 1250 mm.

BAMBOOS OF NEPAL

In Nepal bamboos are abundant between midhills and the terai with most species being found in midhills. Two types of bamboos are commonly recognised and termed as bans and nigalo. The former are of broader diameter, and the latter are of smaller diameter. The nigalo bamboos do not occur in the terai. In the high mountains bamboos are categorized into three types viz. bans, nigalo and malingo. The last named have the smallest diameter. Bamboos are abundant in the eastern, central and western hills, and in the terai region, but not in the midwestern and far western regions of Nepal. Das (1988) has summarised the distribution of some of the useful bamboos of Nepal. The main genera are *Arundinaria, Bambusa, Dendrocalamus, Drepanostachyum* and *Thamnocalamus*.

Two species of Arundinaria occur in Nepal viz. *Arundinaria maling* commonly known as malingo occurs at high altitudes and is widespread over 2800 metres in eastern Nepal. *Arundinaria racemosa* is found above the altitudinal range of *A.maling*.

Bambusa arundinacea commonly known as kante bans, is exotic to Nepal and is cultivated in far-western Nepal. *Bambusa balcooa* known locally as dhanu, bhalu, or harouti bolka bans is a big sized, large diameter bamboo which occur from terai to midhills. *Bambusa nutans* known locally as mal bans is a large sized bamboo and an important species of eastern Nepal. It occurs from terai to 1600 metres. There is an unidentified bamboo species known locally as tharu bans which is large sized, and an important species of central Nepal. It is found in Kathmandu, and Pokhara valley, and in the inner plains and foot hills. Another unidentified species is locally known as mokha bans and is common in terai regions of eastern and central Nepal. Still another unidentified species is known locally as koraincho bans and is commonly found in the inner plains and Chure hills of central Nepal both in the natural forests and farmlands. *Bambusa vulgaris* is an exotic and planted for ornamental purposes.

There are six species of *Dendrocalamus* which occurs in Nepal. *Dendrocalamus hamiltonii* is commonly found in the hills mainly between 300 and 2000 metres and is known locally as tama or choya bans. *Dendrocalamus hookerii* is common in the eastern hills between 1500 to 2000 metres and is known locally as kalobans or balu bans. *Dendrocalamus patellaris* is frequently found in the Mechi hills of eastern Nepal between 1900 and 2600 metres. It is also found in Palpa district and in high rainfall area around Pokhara valley in the western region. It is known by a variety of local names such as nibha, leyas, murali or gopibans. *Dendrocalamus strictus* is found in the terai region below 1000 metres in farm lands, as well as in the natural forests of the mid and farwestern terai. It is an exotic introduced from India and is known locally as kath or lathi bans. An unidentified *Dendrocalamus* species which is known locally as chyoa bans, khosre, tama, or phosre bans resemble *D. hamiltonii* and is found between 1500 to 2000 metres on a large scale. Yet another unidentified *Dendrocalamus* species is know locally as dhungre bans. It is the largest-diameter bamboo of Nepal and is commonly found between 1500 to 2000 metres in central Nepal and less commonly in the eastern hills.

There are five species of *Drepanostachyum* occurring in Nepal out of which three are unidentifed viz. *Drepanostachyum* species known locally as malinge nigalo is an eastern species found between 1800 to 2200 metres. Another species of *Drepanostachyum* also known as malinge nigalo occurs in central Nepal from 1800 to 2800 metres in the natural forest. Yet another species of *Drepanostachyum* also known locally as malinge nigalo occurs in western Nepal and looks similar to the other two above mentioned species. *Drepanostachyum khasianum* known locally as tite nigalo is a species of central and western Nepal and is found in the natural forests. *Drepanostachyum intermedium* also known locally as tite nigalo occurs from 1200 to 2400 metres in natural forests of eastern Nepal.

Four species of *Thamnocalamus* occur in Nepal. *Thamnocalamus spathiflorus* occurs above 2000 metres under deodar and fir forests of far western and mid-western Nepal. It is known locally as ringal. An unidentified species of *Thamnocalamus* known locally as ghoonre nigalo occurs in central Nepal between 2000 and 2500 metres. Another unidentified species which is known locally as chigar occurs in the high mountains of western Nepal. Still another unidentified species of the high mountains of western Nepal is known locally as Jarubuto.

BAMBOOS OF SRI LANKA

Sri Lanka is poor in its bamboo flora and consists of only ten species according to a recent revision of the group by Soderstrom and Ellis (1988). A remarkable feature however is the high degree of endemism with one genus viz. *Davidsea* and eight species being repored as unique to the country (Neela de zoysa et al 1988).

In Sri Lanka bamboos grow in all climatic zones except the very dry areas. *Bambusa bambos* and *Dendrocalamus cinctus* occur in the dry zone of the country and the latter is very restricted in distribution and is known only from one or two isolated forested inselbergs such as the Ritigala located in the north-central region of the country. *Ochlandra stridula* on the other hand is found extensively in the wet lowlands of the south western region. The remaining seven species viz. *Davidsea attenata Pseudoxytenanthera monadelpha, Arundinaria deblis, A.floribunda, A.scandens, A.walkeriana* and *A.densifolia* are found in the high altitudinal montane areas of the central hill country. Most of the bamboo species occur as an understorey. The exceptions are *Dendrocalamus cinctus* and *Arundinaria scandens* which have been reported from windswept mountain tops and *Arundinaria densifolia* from low temperature swamps within the montane grasslands. Apart from these, there are also some introduced species of bamboos such as *Bambusa multiplex* from China, *Dendrocalamus giganteus, D.membranaceous, D.asper, D.strictus,* and *Thyrsostachys siamensis* as well as two varieties of *Bambusa vulgaris.*

BAMBOOS OF BANGLADESH

Nearly 33 species of bamboos occur in Bangladesh in the north, the east, and the south, especially Chittagong hill tracts. About 900 square kilometers of reserved forests of Sylhet and Chittagong hill tracts can be classed as bamboo forests (Sharma 1980).

BAMBOOS OF BURMA

In Burma 90 species have been recorded and occur all over the country in the nine forest types either as an understorey or as a pure stand. *Melocanna baccifera* for instance occurs as pure stands over 7800 square kilomoters in Arakaan Yoma (Sharma 1980). Only seventeen genera are utilized commercially viz. *Sinobambusa, Chimonobambusa, Arundinaria, Phyllostachys, Bambusa, Thyrsostachys, Gigantochloa, Dinochloa, Oxytenanthera, Dendrocalamus, Dendrochloa, Pseudostachyum, Schizostachyum, Neohouzeoua, Cephalostachyum, Melocanna and Teinostachyum.*

BAMBOOS OF THAILAND

In Thailand a study of bamboo taxonomy is believed to be still in its initial stages. However it is estimated that there are 12 genera and 41 to 50 species. Bamboos occur widely in the north, northwest, and south, in evergreen and deciduous forests in a contiguous belt with Burma. A number of yet unidentified species are expected to be found in the Tenasserim mountain range along the border of Burma, and the Kalakhiri range along the border of Malaysia.

Distribution of bamboo species varies with the climate, soil fertility, and elevation, although some species such as *Bambusa arundinacea, Thyrsostachys siamensis* and *Gigantochloa albociliata* have very wide distribution. Some species are very restricted in their distribution and occur only in particular areas such as *Schizostachyum zollingeri* steud and *Gigantochloa hasskarliana* are confined to the peninsula in the tropical rain forest. *Dendrocalamus giganteus, D.hamiltonii, D.sericeus Munro, Melocalamus compactiflorus* Benth., and *Teinostachyum griffithii* Munro occur in the hills of the north, while *Bambusa arundinacea* occurs along river banks and water courses in the lowland, and *Bambusa tulda* and *Thyrsostachys oliveri* prefer the deep soil in the valley where the relative humidity is high. The Arundinarias occur in the dry deciduous dipterocarp forests e.g. *Arundinaria ciliata, A.camus* and *A.pusilla A.chev* where they are preponderant.

The lesser known species present only as herbarium sheets are (i) *Dendrocalamus dumosus* (Ridley) Holttum from Rawai Island Satun (ii) *Bambusa pierrei* E.G.Camus, (iii) *Bambusa wamin* Brandis from ChiangMai, (iv) *Oxytenanthera stocksii* Munro from Nakhon Phanom, (v) *O.hosseusii Pilger* from Nakhon Thai, Phitsanulok, (vi) *Schizostachyum insulare* Ridley from Rawai Island, Satun, and (vii) *S.longispiculatum* Kurz. Further, some of the bamboo species found in aerial survey and ground check in Thailand are (i) *Thyrsostachys siamensis* in the central region, (ii) *Bambusa arundinacea* in Kanchanaburi, (iii) *Dendrocalamus longispathus* in Khao Hin Lap and (iv) *D. strictus* in Erawan Srisawat.

In the northern Lampang Ngao area *Thyrsostachys oliveri, T.siamensis, Bambusa arundinacea, B.blumeana, Gigantochloa albociliata, Bambusa tulda, Dendrocalamus membranaceous* and *Schizostachyum aciculare* occur.

In the southern region in Surat Thani are *Oxytenanthera hosseusii, Gigantochloa nigrociliata, Bambusa arundinacea, Schizostachyum aciculare* and *S.zollingeri.*

In addition to the above mentioned species there are also some introduced species cultivated for handicrafts and other industries such as *Bambusa polymorpha, B.nana, Dendrocalamus asper* Back, *D.brandisi* Kurz, *Cephalostachyum pergracile* Munro, *C.virgatum* Kurz and *Gigantochloa macrostachya* Kurz.

BAMBOOS OF MALAYSIA

In Malaysia bamboos occur gregariously in localized patches, on river banks, on hill sides, ridge tops and in disturbed lowland forests. Bamboos occur in abundance in areas with a history of shifting agriculture or large scale logging e.g.. in the Tapah district of Perak, the Ulu Langat district of Selangor and the Mata Ayer district of Perlis, in Sabah, and in the Ma'okil Forest Reserve of Johore. Genera of bamoos that occur in Malaya are *Bambusa, Dendrocalamus, Dinochloa, Gigantochloa and Schizostachyum.*

Ng and Shamsuddin 1980 have summarised the occurrence of different species of bamboos in Malaysia. There are fourteen species of *Bambusa* in Malaya of which two are originally from China and Japan viz. *Bambusa ventricosa* and *B. glaucescens*. The rest of the species are *Bambusa arundinacea* which occur in Penang and Singapore botanic gardens, *B.blumeanā* occurs in Penang, Selangor (Kepong) and Pahang (Pekan). *Bambusa burmanica* occurs in Kedah, Alor Star and Singapore. *B.heterostachya* occurs in Perak, north Sembilan, Meloka, and Johore. *B.klossii* occurs in Kedah, and Perak at about 1000 metres, *B. magica* occurs in Pahang, Cameron Highlands, Selangor, and Ulu Semangkok. *B.montana* occurs in Penang Hill and Kedah; *B.pauciflora* occurs in Pahang and Singapore. *B.vulgaris* occurs all over Malaysia. *B.wrayi* occurs in Perak-Gunong Inas at 1500 to 2000 metres.

There are eight species of *Dendrocalamus* viz. *Dendrocalamus asper* is cultivated for its shoots all over Malaysia, *D.dumosus* occurs in Kedah (Baling Hill) and Langkawi, *D.elegans* occurs in Dedah, Pulau Zangkawi, and Penang. *B.hirtellus* occurs in Johore, Perak. (Taiping), Kedah, and Kelantan. *D.pendulus* occurs in Perak, Selangor, and north Sembilan, *D.sinuatus* occurs in Perak, north Sembilan, Trengganu, and Pahang. Apart from these there are two introduced species viz. *Dendrocalamus giganteus* from Burma and *D.strictus* from India.

Eight species of *Gigantochloa* occur in Malaya viz. *G.apus* occurs in Selangar (Serdang) and Singapore botanic gardens, *G.hasskarliana* occurs in Penang and Singapore, *G.latifolia* occurs in Kedah, Perak and Pahang. *G.levis* occurs in Selangor, Melaka, Johore and Singapore, G.ligulata occurs in Perlis, Kedah, Perak, Pahang, Selangor, and Kelantan. *G.maxima* has three varieties viz. variety *viridis* occurs in Johore and Kota Tinggi, variety *minor* occurs in FRI Kepong, variety ridley occurs in province Wellesly and Singapore botanic gardens. *G.scortechenii* occurs in Kedah, Penang, Perak, Selangor, north Sembilan, Kelantan and Pahang. *G.wrayi* occurs in Kedah, Pahang. province Wellesly, Perak and Selangor.

Nine species of *Schizostachyum* occur in Malaysia viz. *Schizostachyum aciculare* occur in Johore, Selangor, north Sembilan, and Melaka. *S.brachycladum* occurs in Kedah, Penang, Perak, Pahang and Johore. *S.gracile* occurs in Johore, Selat Teberau, Kota Tinggi, Segamat, Sg. Sedili, Melaka, Bukit Tungal, Air Panas, Selangor, Sungai Labu, Pahang, Kuala Bera and Pekan *S.grande* occurs in Kedah, Grik, Perak, Cameron Highlands, Pahang, Kuala Lipis, Kelantan and Selangor. *S.insulare* occurs in Pulau Lankawi, Kedah, Penang and Johore. *S.jaculans* occurs in Pahang, Selangor, Melaka, Johore, Singapore and Perak. *S.longispiculatum* occurs in all states except Perlis, Kedah, Pinang Trengganu, province Wellesley. *S.terminale* occurs in Kedah, Inchong Estate, Sg.Kenan. *S.zollingeri* occurs in Perak Selangor, north Sembilan, Pahang and Johore.

BAMBOOS OF INDONESIA

In many areas of Indonesia, bamboos are a cheap and plentiful resource for the vast and varied needs of the people. But because of the lack of information on bamboo taxonomy of Indonesia, data on distribution of individual species are also lacking. Backer in 1924 had listed 31 species of bamboos in his Handbook voor de Flora van Java. Since then, some studies taken up have revealed that a larger number of species is likely. Based on the presence of large tracts of land over grown by bamboo in Gowa and Maros districts in south Celebes, the Government installed a paper factory. A similar factory was also installed in east Java. The bamboo species commonly used in Indonesia for various purposes were listed by Widjaja in 1980 and are as follows :- *Arundinaria japonica, Bambusa arundinacea, B.atra, B.blumeana, B.glaucescens, B.polymorpha, B.vulgaris, Dendrocalamus asper, Dinochloa scandens, Gigantochloa apus, G.atter, G.verticillata, Nastus elegantissimus, Phyllostachys aurea, Phyllostachys nigra, Schizostachyum blumei, S.brachycladum S.caudatum, S.lima, S.zollingeri* and *Thyrsostachys siamensis.*

BAMBOOS OF THE PHILIPPINES

About 55 species of bamboos have been recorded in the Philippines under ten genera viz. *Bambusa, Dendrocalamus, Gigantochloa, Yushania, Schizostachyum, Thyrsostachys, Leleba, Phyllostachys, Cephalostachyum,* and *Dinochloa*. Out of the 55 species about 36 are erect bamboos and the rest are climbing.

The species that has the most extensive distribution throughout the country is *Bambusa blumeana*, Schultes f. It occurs along the side of rivers and creeks and is extensively cultivated. A forest of *Bambusa blumeana* occurs in Cardona Rizal. In general, the genera *Bambusa, Schizostachyum* and *Dendrocalamus* are scattered from north to south in the islands. Only one species of *Phyllostachys* viz. *P.nigra* var *henonis* (Mitf) stapf. ex. Rendle grows favourably in Philippines. Depending on the species, bamboos occur from sea level at the coasts, to an altitude of 2800 metres and occasionally upto 3200 metres. They thrive best on well drained sandy loam to clayey loam derived from river alluvium or from underlying rocks having a pH of abut 5.0 to 6.5. Large tracts of bamboos occur in the northern provinces on marginal lands, courses of streams, and rivers, village homelots, and hillsides. Several climbing species of bamboos such as *Dinochloa* species form dense thickets in the forests in the southern region.

Eleven speices of *Bambusa* occur in Philippines viz. *Bambusa arundinacea* willd introduced from India, *B.floribunda* Nakai, *B.blumeana* Schultes f. known locally as kanayan-tinik, *B.cornuta* Munro known locally as lopa, *B.merrillii* Gamble, *B.nana Roxb; B.tulda Roxb, B.vulgaris* Schrad ex. Wendil known locally as kauayan-killing, *B.vulgaris* variety striata (Lodd) Gamble, *B.ventricosa* McClure, and *B.multiplex* (Lour) Raeusch.

Dendrocalamus occur as four species viz. *Dendrocalamus merrillianus* (Elm) Elm; *D.curranii* Gamble, *D.latiflorus* Munro and *D.parviflorus* Hack. *Gigantochloa* occurs as *Gigantochloa aspera* Kurz; *G.levis* (Blanco) Merr, *G.scribneriana* Merrill. One species of Guadua occurs as *Guadua philippinensis* Gamble, two species of Leleba occurs as *Leleba floribunda* Nakai and *L.ventricosa* McClure and one speices of Thyrsostachys viz. *T.siamensis* Gamble.

Apart from *Phyllostachys nigra* some other species of phyllostachys have been introduced to Philippines viz. *P.bambusoides* var *aurea* Makino, *P.pubescens* Mazel ex-H Lahaie, *P.aurea* carr, *P.edulis* Makino. Six species of erect Schizostachyum occur in Philippines viz. *Schizostachyum lima* (Blanco) Merr. *S.brachycladum* Kurz, *S.hirtiflorum* Hack, *S.lumampao* (Blanco) Merr, *S.textorium* (Blanco) Merr and *S.zollingeri* steud. The other species of the same genus are climbing bamboos viz. *Schizostachyum acutiflorum Munro, S.curranii* Gamble *S.dielsianum* (Pilger) Merr, *S.luzonicum* Gamble, *S.mucronatum Hack, S.merrilli* Gamble, *S.palawanese* Gamble, *S.toppingii* Gamble.

Other climbing bamboo species include *Dinochloa aquilarii* Gamble, *D.ciliata* Kurz, *D.elmeri* Gamble, *D.luconiae* (Munro) Merr, *D.pubiramea* (merr) Gamble, *D.scandens O.Kuntze, D.scandens* Var *angustifolia* Hackel ex Merr and *Cephalostachyum mindorense* Gamble.

BAMBOOS OF JAPAN

Studies on the morphology and distribution of bamboos in Japan have been going on since 1900, and during nineteen thirties bamboo distribution in various districts of Japan was investigated by Nakal (1933-1935) and Koidzumi (1934 to 1943) and they discovered a large number of new species. The number at present is estimated to be 13 genera, 669 species, 23 varieties and 16 forms (Oye et al 1980). Suzuki in 1978 prepared an index to Japanese bamboos and he classified them into 13 genera, 8 sections, 95 species, 7 subspecies, 63 varieties, and 72 forms. The two main species of Japanese bamboos are *Phyllostachys reticulata* C.koch also known as *P.bambusoides* Sieb (Local name Madake) and *P.edulis* Makino locally known as *Mosochiku.* These two species occupy 84 per cent of the total area occupied by bamboos. Another important species is *Sasa* which is a bamboo like plant growing mainly in Japan and has many species. It is difficult to distinguish Sasa from bamboo although it is generally smaller than bamboos and the culmsheaths are persistent unlike in the bamboos where they are deciduous. The genus Sasa is divided into *Macrochlamys, Sasa* (i.e. Eusasa). *Lasioderma, Grassinodi* and *Moniclade.*

Macrochlamys and *Sasa* grow mainly in the northeastern district of Japan. The most important species is *Sasa kurilensis* (Ruprecht) Makino et Shibata variety *kurilensis.* It is abundant in heavy snowfall districts in Hokkaido and Tohoku and is also found in Hokuriku, San-in, central and southern Saghalien, Kuriles, and northeastern districts of the Korea Peninsula.

BAMBOOS OF PAPUA NEWGUINEA

Twenty six species of bamboos have been reported from Papua New Guinea and they occur extensively in the Savannah of the western province. Many villages in the lowlands and highlands have been planted with thick walled bamboos for construction purposes. The thick walled *Bambusa vulgaris* is being increasingly used for furniture, novelties, decoration and handicrafts. The thin walled *Schizostachyum lima* is woven into sheets and used as cladding for walls in houses. A "bambusarium" has been established at Lae and in the Agricultural Experimental Station at Laloki (Port Moresby) with species from south east Asia and south America (Sharma 1980).

BAMBOOS OF AFRICA

About 14 genera and 43 species of bamboos have been reported to occur in Africa all of which are mainly distributed in the east African region. There has been no proper inventory of bamboo resources in Africa but 43 species have been documented as being from east Africa (Kigomo 1988). The more widespread genera are

Arundinaria, Oreobamboos, and *Oxytenanthera*. The rest of the eleven genera occur mainly in Madagascar and include genera Cephalostachyum, *Decaryochloa* Hickella, *Hitcheckella, Nastus, Perrierbambus, Pseudocoix*, and *Schizostachyum*. Distribution of patches of bamboo in east Africa occurs from temperate highland forest zone of Ethiopia and upper reaches of Nile River in the north, to the Basutoland Highlands, Natal and Madagascar in the south. The east African bamboo species form vast pure stands of single species or are in association with other trees as an understorey.

Arundinaria alpina K.Schum is distributed between 2290 to 3360 metres above sea level. It occurs gregariously within mountain forest in tropical Africa and is found growing in the highlands of Ethiopia, southern Sudan, Congo, Zaire, Rwanda, Uganda, Kenya, Tanzania and Zambia. A species of *Arundinaria* was collected as far south as Cape Province, Natal but this was not confirmed as *A.alpina*. Sterile specimens resembling *A.alpina* have been collected in Malawi and they occur as scattered clumps in broad leaved montane forest formations. Clayton (1970) also recorded the occurrence of A.alpina in Cameroun suggesting this to be its most westerly natural occurrence. In Ethiopia this species occurs in the highlands, in Kenya it occurs in irregular patches in the central highlands, particularly in Timboroa plateau and Aberdare Range. In Mount Kenya it occurs in Elgon and Mau range. In Uganda it occurs in Mbulu, Arusha and Mbeya districts on the highlands of Iringa, Lukwangule and Ulugurus and Mount Meru.

The other important species is *Oxytenanthera abyssinica* (A. Rich) Munro which occurs in open areas in forests and often by rivers at altitudes between 1100 and 2100 metres. It occurs in Ethiopia in the north to Malawi, Zambia and Zimbabwe in the south. In Burundi the species occurs around Bunjumbura and Lake Mossor regions at about 1600 metres and 1400 metres respectively. In Ethiopia its occurrence is on the hillside and savanna woodlands. It is the most hardy of the three commonly occurring east African bamboo species. In Tanzania the species is found on poor soils in dry forest formations. *Oxytenanthera abyssinica* has a wider distribution in Lindi, Kigoma and Usaramo regions. In Uganda it occurs mainly in western Nile, Acholi, Karamoja, and Mound Elgon regions, In Malawi, Zambia and Zimbabwe it is common in semi deciduous dry forest formation.

Oreobambos buchwaldii K.Schum is yet another important bamboo species of east Africa and occurs mainly between 300 and 1930 metres. In Burundi the species occurs in a patch along river Kitima in Bubanza region. In Malawi and Zambia it occurs widely between 400 and 1950 metres. It is more common in open areas along rivers in the forest patches of the Shire highlands in Malawi. In Tanzania it occurs between 450 and 1000 metres in solitary clumps, and in more open parts of the evergreen forests of the east Usambaras and Tukuyu highlands. It is also reported to occur in forest swamps around Mengo, Masaka, Bunyoro and scantily in Busonga forest districts.

In addition to the above three species a number of bamboo species have also been introduced from India, Burma, Thailand, China and Japan. These include species of Arundinaria, Bambusa, Chimonobambusa, Dendrocalamus, Gigantochloa, Melocalamus, Phyllostachys, Phragmites and Shibatea.

PATTERNS OF GEOGRAPHICAL DISTRIBUTION

Ecology and geography of plants are of much relevance to plant taxonomy because each taxon exhibits a certain pattern of distribution which is one aspect of its definition. Areas occupied by related taxa has a bearing on the classification of the group and it becomes of special significance when the evolution of the species is taken into consideration. Geographical patterns also become significant because plants are collected from identified areas for cataloguing in herbaria and describing in floras. The concept of geographical distribution and ecology form the subjects known as phytosociology, phytogeography and plant genecology. Geographical differentiation exists between taxa at all levels and in all degrees of spatial separation because different taxa may possess quite different powers of migration. Taxa which occupy mutually exclusive geographical areas are termed allopatric and those occupying similar or overlapping areas are termed sympatric. Closely related sympatric taxa usually show different types of genetic, ecological and structural differentiation from those shown by closely related allopatric taxa, because allopatry is itself an important isolating mechanism. If the distribution of a large number of taxa is analysed within a given area, certain geographical patterns are found to recur consistently. These patterns and the taxa which exhibit them are known as floristic elements. A knowledge of floristic elements is helpful in understanding the pathways of migration and may also aid in taxonomic decisions in cases of doubt (Stace 1989).

If the distribution of every species within a genus or every genus within a family is indicated on a single map, it will usually be found that there are one or more areas with a marked concentration of the species or genus as the case may be. Such an area is termed as a 'centre of genetic diversity' for that genus. Often there is a single such area for the genus concerned, and there is progressively fewer species found as the distance from the centre of diversity is increased. (Stace 1989).

The topic of centres of genetic diversity was first studied for crop plants particularly cereals and legumes by Vavilov in the nineteen twenties and thirties, and an updated discussion of the topic is provided by Zeven and Zhukovsky 1975. Vavilov and later researchers found moreover that there is a frequent coincidence in the centres of diversity for many different unrelated taxa and that a relatively small number of major centres of diversity can be recognized in the world. Zeven and Zhukovsky defined twelve such centres and all of them lie in the tropical, subtropical, and warm temperate areas. Vavilov believed that the centres of diversity were also the centres of origin of the taxa concerned.

Stace (1989) states that each monophyletic taxon has a single centre of origin. By studying the taxon in its centre of diversity in relation to its characteristics, progressively further from the centre, the major patterns of evolution and pathways of migration can very often be deduced. The ways in which the taxon had adapted to the different environments which it has encountered during its migrations are thus frequently uncovered. Such an outward pattern of evolutionary migration is known as 'adaptive radiation'. Its demonstration will be of great assistance in deciding upon the taxonomic relationships and delimitations of various biotypes encountered within a taxon. Frequently, cytological studies also have to be done, as it has been found that plants near the centre of diversity are diploids, while those at the edge of the range are polyploids. Sometimes there is more than one centre of genetic diversity for a taxon in which case it is an indication that the taxon is polyphyletic, having arisen in more than one area. However all but one of these centres are secondary centres of genetic diversity, i.e. areas where the taxon was able to diversify to a greater extent than from the primary centre perhaps due to the presence of a large number of habitats. The decrease in variability away from the centre of diversity is known as "biotype depletion" which also can be used to trace pathways of migration. The concept of centres of genetic diversity and centres of origin has been strongly criticised in recent years by vicariance biogeographers who prefer to think in terms of the "generalised track" of a taxon differentiating allopatrically in different regions. But Stace (1989) believes that something between the two may be more accurate.

GROWTH AND DEVELOPMENT

Bamboos have a finite life span and must periodically go through the entire development sequence starting from germination of seed to development of rhizomes and production of mature culms to finally their flowering and seeding. An understanding of the major features of this development process is important to an understanding of their performance and distribution.

The environmental stimuli to seed germination are varied and include moisture, temperature, light, and atmospheric gases. Wherever the appropriate stimulus for seed germination does not occur, the species will be effectively excluded from establishing. In the intermediate post germination period also, the small size and delicacy of the seedlings at this stage makes them particularly susceptible to environmental stresses such as extremes of temperature and moisture variations. Survivors eventually enter a mature phase in which rhizomes are formed, and culms are produced every year, and finally reproduction is achieved. All these processes have in each species their own specialized environmental requirements, which if unfulfilled will preclude continuance of the species beyond one generation. The final senescent stage of the life cycle is an ill defined and poorly understood phase during which there is a complete breakdown in the overall activities of the clump.

Some generas of bamboos are monopodial eg. *Phyllostachys* and produce widely separated culms, while others such as *Bambusa* and *Dendrocalamus* are sympodial and clump forming, and still others such as *Dinochloa* are climbing. There are some species which are intermediate between monopodial and sympodial with somewhat open clumps eg. *Dendrocalamus membranaceous*. The monopodial bamboos are characteristic of subtropical and temperate regions, and sympodial bamboos are characteristic of tropical countries.

The full length in height in all bamboos is reached in the first season. For about a year the culms are soft but gradually within the next two years they become hard by deposition of lignin in the fibres and silica in the parenchyma and epidermis. The culms of different seasons are distinguished by the presence or absence of culm sheaths and a change in the colour and texture of the culm epidermis. The number of culms in a clump varies with the species, and to a limited extent within the species itself. But no detail work has been done on this aspect.

The flowering rhythm of bamboos in general have been described very aptly by Gaur (1985) as well as a lot of other researchers earlier. The rhythm varies in different species and range from annual flowering in *Bambusa atra* to constant sterility in *Bambusa vulgaris.* Majority of bamboos fall between these two extremes and the flowering cycle ranges from annual flowering, to once in a life time, The life span again varies with the species and may extend from 7 years in some species to 60 years in others. Some species such as *Bambusa atra* and *Ochlandra acriptoria* also known as *Ochlandra rheedii* flower annually but *not* gregariously. Then there are species that flower periodically and gregariously such as *Bambusa arundinacea* with a life span of 30 to 45 years and *Dendrocalamus strictus* with a life span of 20 to 60 years. There are also species such as *Bambusa polymorpha* that flower sporadically and irregularly and has a life span of 30 to 60 years. In most cases the culms, and in gregariously flowering species, the whole clumps, die after flowering. Those that flower annually do not die. Another aspect of flowering is the synchronous flowering. An example given by Gaur 1985 is that of *Thyrsostachys oliveri.* This species flowered and seeded in Burma in 1891, and seeds were planted in Calcutta and Dehradun about 1500 Km apart. The clumps at both places flowered synchronously in 1940. Such synchronised flowering was also observed in *Melocanna baccifera* known also as *Melocanna bambusoides* in Garo hills (Assam) and Dehradun (Uttar Pradesh). Such physiological processes as flowering that measure time by some means, and do so in relation to environmental time cues is known as biological clocks. The study of biological clocks in general was scarcely considered scientific a few decades ago. But now it is one of the active growth points in biology, cutting across ecology, ethology, physiology, biochemistry, biophysics and cybernetics. Recognition of the importance of biological timing for ecology, behaviour, and physiology has become of widespread interest (Brady 1979).

The ontogeny of bamboos thus poses a sequence of environmental requirements which includes both environmental resources utilized in growth and development, and other non resource environmental conditions that affects its physiology at some stage of its life cycle. The seed phase especially of the life cycle, represents a phenomenon of special geographical interest as the seed represents the means whereby one plant of the finite life span, creates a new plant and ultimately ensures perpetuation of the species. Bamboos vary widely from species to species in their seed production. Although this variability may be to a certain extent environmentally induced, it is to a large extent, internally controlled when mass seeding occurs synchronously at intervals of several decades. Janzen in 1976 suggested that this mass seeding may represent a strategy for saturating seed predators that would otherwise consume all of a continuously produced seed crop.

In any case plant populations rarely experience the opportunities for unrestricted growth, except perhaps for a period of time. Inevitably the population approaches a limit to further growth imposed by the carrying capacity of the environment whose effect may be mediated by interspecific of competition. Plants respond to this crowding by (i) lower birth rates, (ii) slower maturation, (iii) smaller size and (iv) higher death rates. This may explain why bamboos flower and seed at long intervals, although they do multiply to a limited extent by rhizomes. This may also explain why bamboos grow to their full height in one season and why they do not expand laterally as trees do.

4

MORPHOLOGY AND ANATOMY

Bamboos as already discussed so far, have been very successful in colonizing a very wide range of habitats extending from cold temperate regions to moist tropical forests, and from sea level to high altitudes. Correspondingly they also exhibit an astonishing range in size, form, and structure, as evident from the genus *Sasa* of Japan at one end of the scale, and *Dendrocalamus giganteus* of Burma at the other end of the scale. The range of form and structure in bamboos have been studied by different researchers at different stages of their life cycle depending upon the availability of material. We therefore have scattered piecemeal information on different speices available in reports, research papers, floras, and a few books on bamboos. In order to get these information into their correct perspective, it is necessary to present a coordinated whole picture of the group.

Life cycle of bamboos start with germination of the seed. In the caespitose type with grain like seeds, when germination takes place, the plumule emerges in the form of a pointed conical bud with sheathing scale like leaves. It rapidly develops into a thin wiry stem bearing single linear foliage leaves arising alternately at the nodes, with the bases of the leaves sheathing the stem. Meanwhile fibrous roots develop from the base of the young shoot. The tufted form of the young plant commences to show when the fibrous roots develop successive pointed buds which develop into the rhizome. These rhizomes curve upwards and form aerial shoots. The buds, rhizomes, and the shoots arising from them gradually increase in size, season after season. The earlier shoots are thin, wiry and grass-like but subsequently after four to five years the enlarged rhizomes produce woody culms surrounded by culm sheaths, and a number of such culms form the clump. A mature bamboo therefore consists of a woody rhizome from which are developed woody culms divided into prominent internodes and nodes, and each internode is at the early stages covered by a culm sheath which varies in morphological structure from species to species. Branches develop at the nodes from dormant buds, a year after the culms have attained their full height.

RHIZOMES

McClure (1966) gives a detail account of bamboo rhizomes. There are two types of rhizomes (i) the monopodial also termed as dumetose, and leptomorph by different researchers (ii) The sympodial also referred to as caespitose and sometimes as pachymorph. The monopodial rhizomes produce culms 30 to 90 cm apart, while sympodial rhizomes produce closely packed culms which from clumps (Pl 9 fig. 1&2)

The bamboo rhizome is an underground stem and in some ways similar to the culm in structure. It consists of compressed internodes and closely spaced nodes from which the roots develop. The buds on the rhizome are initially flat in shape and usually less than 2.5 cm in diameter. The rhizome itself is covered profusely with scales which at first are not very apparent, but develop rapidly as the buds develop.

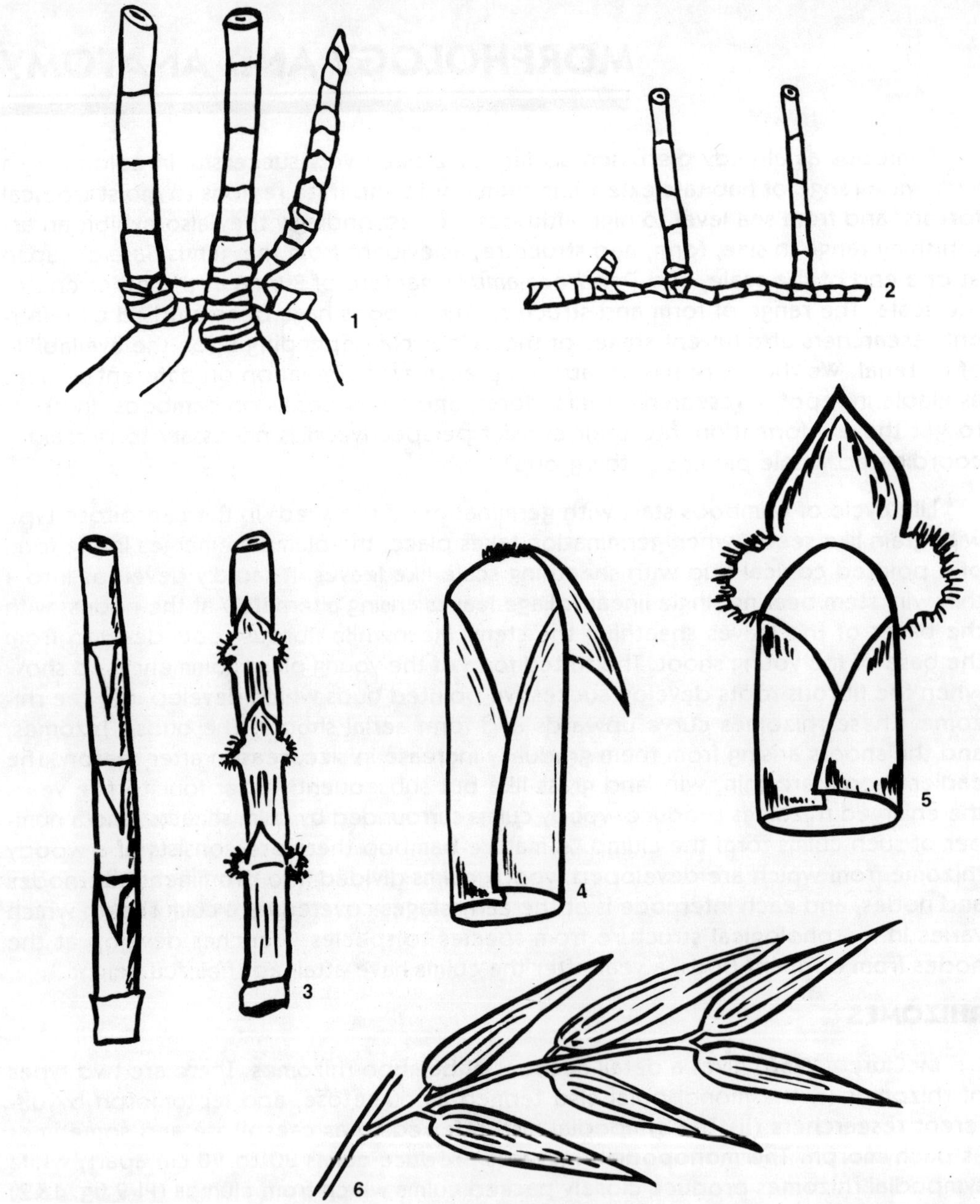

Pl 9. Morphology of Bamboo Culms. 1. Caespitose type rhizome of *Bambusa polymorpha* 2. Dumetose type rhizome of *Melocanna bambusoides* 3. Young culm portions of *Dendrocalamus longispathus* and *B. tulda* 4&5. Culm sheaths of *D. longispathus* and *B. tulda*. 6. Leaf of *D. longispathus*.

These scales are rudimentary sheaths. On careful dissection of the scales, and the bud, it is seen to contain a complete bamboo in embryo, and may consist of a number of telescopic internodes with an equal number of scales with the appearance of a terraced mound. All buds do not develop into culms, but those that do, are usually one or two years old, seldom more. They start their growth with the beginning of the rains and appear above the ground soon after. During the development of a rhizome bud into a culm, the first thing that is formed is a short new rhizome. Before the culm appears above ground, the new growth develops into a complete rhizome for the support of the culm. At this stage, it is equipped with several fully grown buds, which in their turn lie dormant until the following years.

The function of the rhizome is to give support to the growing culms and to act as a feeding channel. The feeding organs of a rhizome are the slender rootlets which are capable of stretching out from a few centimeters to a meter or so deep into the ground. The sheaths are of vital importance to culms during and before growth. Each internode of the new culm is carefully wrapped up in a single sheath, but the basal internode always has more than one sheath. In a bud, the successive sheaths are arranged alternately clockwise and anticlockwise, and their function is to protect the tender internodes against injury and dessication. The outer surface of the sheath is armed with a cluster of stiff hairs which are detatched at the slightest touch and cause irritation.

CULM

The culm is the aerial shoot developed from the rhizome each year. In the initial stages it appears as a cone wrapped up in a series of culm sheaths of decreasing sizes towards the tip of the cone. There is *no* terminal growth. The height growth of the culm is caused by successive elongation of the internodes. Several internodes from the bottom to the top grow simultaneously. The daily increase in length of the internodes during this period ranges from 1.5 cm to 7.5 cm. Completion of elongation of the internodes takes place within a few days. During the time of this growth in height of the culms, the internodes are covered by the culm sheaths which elongate along with their respective internodes (Pl 9 fig. 3). Although the culms grow to their full height in two to three months, they are immature till about three years. During the three years required for maturity, the culms gradually lose their culm sheaths, decrease in moisture content, and increase in lignin content. Sun Chengzhi and Xie Guoen (1985) found that there is also a relative decrease in crystallinity with age of the culm. The crystallinity was measured by them by crushing samples of the culm into powder with 80 to 100 mesh. The powder was then dried over phosphorous pentoxide in vacuum at room temperature. 500 mg of the powder was pressed into a tablet and used for X-ray measurement for crystallinity.

MORPHOLOGY OF THE CULM

After the culms have attained their full length, the buds which arise in the axils of the sheaths, develop in the upper portion of the culm into leaf bearing jointed woody branches which generally stand in half whorls. In many species such as *Thyrsosotachys oliveri, Dendrocalamus hamiltonii, Melocanna bambusoides* and *Bambusa*

Table 6. Morphology of the Culm of the Some Species of Bamboos

S. No.	Name of Species	Total height of culm in metres.	Total no. of inter nodes	Range in length of internodes in cm.	Range in diameter of internodes in cm.	Range in wall thickness of internodes in cm.	Branching of culm
1.	Bambusa arundinacea	13.95	48	(16.0-37.3)	(1.0-4.7)	(0.45-1.15)	Branches present at lower internodes.
2.	B. nutans	14.17	58	(14.0-40.6)	(0.5-5.2)	(0.15-1.45)	Branches present at lower internodes.
3.	B. polymorpha	13.37	34	(6.0-77.0)	(2.4-7.5)	(0.3-2.15)	Restricted to higher internodes.
4.	B. tulda	7.44	24	(12.0-50.0)	(0.5-5.3)	(0.25-2.65)	Restricted to higher internodes.
5.	Cephalostachyum pergracile	23.77	32	(13.0-82.3)	(0.5-5.6)	(0.15-1.6)	Restricted to higher internodes.
6.	Dendrocalamus hamiltonii	15.74	45	(11.0-46.5)	(0.3-7.5)	(0.1-1.9)	Restricted to higher internodes.
7.	D. Longispathus	11.16	30	(10.0-50.8)	(0.5-6.3)	(0.15-1.95)	Restricted to higher internodes.
8.	D. Strictus	11.02	43	(10.5-38.0)	(0.3-5.2)	(0.15-2.6)	Restricted to higher internodes.
9.	Melconna bambusoides	8.55	31	(8.0-37.8)	(0.5-3.5)	(0.1-1.5)	Restricted to higher internodes.
10.	Oxytenanthera abyssinica	3.18	26	(15.5-27.8)	(0.3-3.4)	(0.1-1.7)	Restricted to higher internodes
11.	O. nigrociliata	10.72	32	(1[illegible].7-43.5)	(1.0-5.4)	(0.15-1.7)	Restricted to higher internodes.
12.	Thyrsostachys oliveri	12.24	38	(14.3-56.7)	(0.4-5.2)	(0.2-3.6)	Restricted to higher internodes.
13.	Dinochloa maclelandi	4.07	11	(19.0-49.5)	(1.3-2.15)	(1.07-1.3)	Restricted to higher internodes.

(Source : Pattanath 1965)

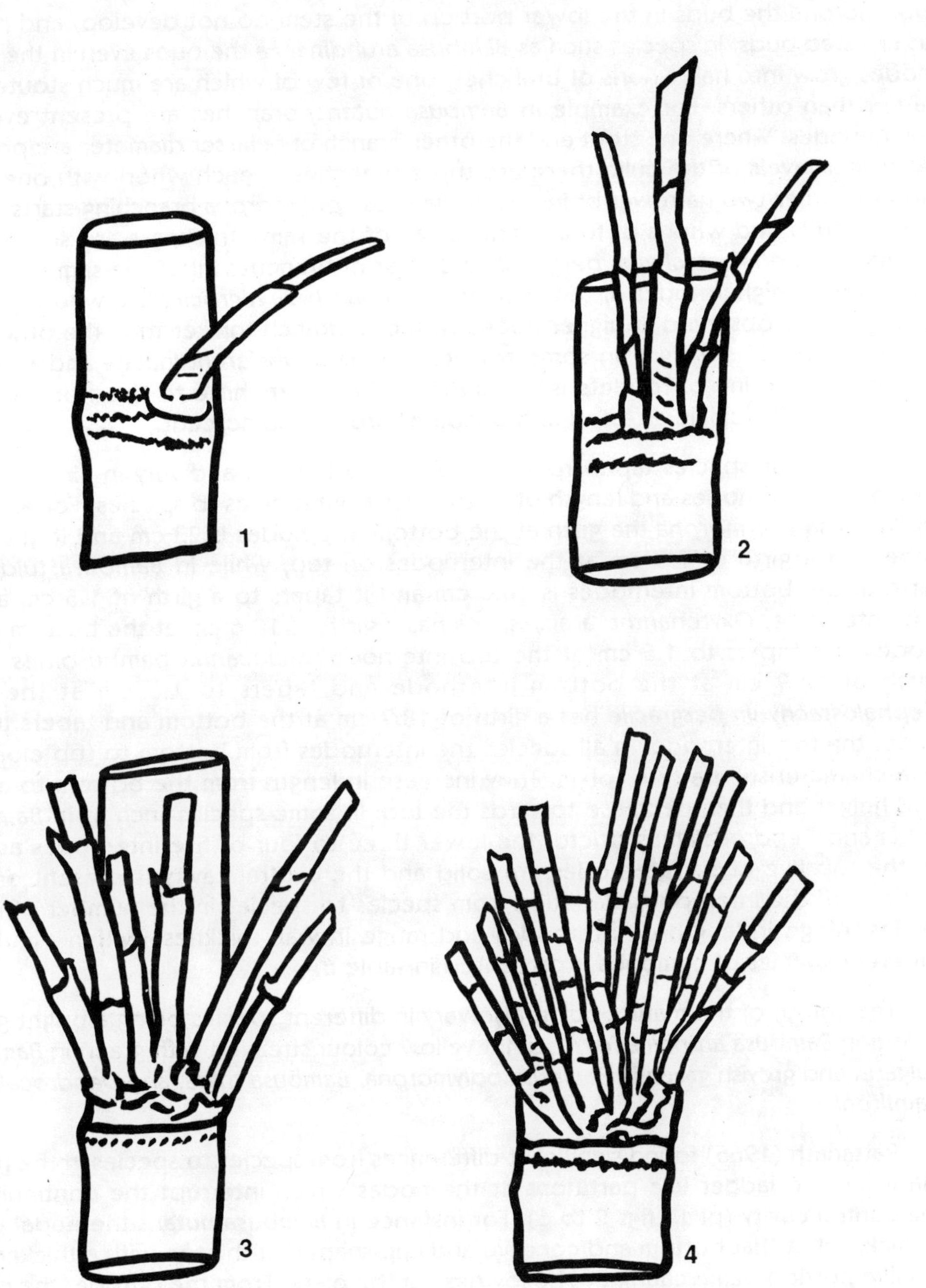

Pl 10. Branching pattern in different bamboo species. 1. Bambusa arundinacea 2. *Bambusa nutans* 3. *Bambusa polymorpha* 4. *Cephalostachyum pergracile.*

polymorpha the buds in the lower portion of the stem do not develop, and remain as arrested buds. In species such as *Bambusa arundinacea* the buds even in the lower nodes grow into half whorls of branches, one or few of which are much stouter and larger than others. For example in *Bambusa nutans,* branches are present even on lower nodes. Where one stout and the other branch of a lesser diameter are present. At higher levels of the culm there are three branches in each whorl with one stout and the other two narrower branches. In *Bambusa polymorpha* branching starts higher up the culm and whorls of four branches all of the same thickness are seen at the nodes. In *Cephalostachyum pergracile* a cluster of branches all of the same girth are observed at higher nodes of the culm. In *Oxytenanthera nigrociliata* a whorl of three branches are observed at higher nodes with one branch thicker than the other two in the whorl (Pl. 10 Figs). In some species e.g. *Bambusa arundinacea* and *Bambusa polymorpha* a ring of rootlets is observed in the lower three to five nodes. These rootlets mostly remain arrested and become hard and spinescent.

Culms in all species taper from the bottom to the top and vary in girth and the number of internodes and length of internodes from species to species. For example in *Bambusa polymorpha* the girth at the bottom internodes is 23 cm and it gradually tapers to a girth of 7.5 cm at the internodes on top, while in *Bambusa tulda* the girth at the bottom internodes is 16.6 cm and it tapers to a girth of 1.5 cm at the top internode. *Oxytenanthera abyssinica* has a girth of 10.6 cm at the bottom internodes and tapers to 1.5 cm at the top internode. *Melocanna bambusoides* has a girth of 10.9 cm at the bottom internode and tapers to 1.5 cm at the top. *Cephalostachyum pergracile* has a girth of 18.7 cm at the bottom and tapers to 1.2 cm at the top internode. In all species the internodes from bottom to top elongate in a characteristic pattern (pl 1). They increase in length from the bottom to about mid height and then decrease towards the top. In some species such as in *Bambusa tulda* and *Dendrocalamus strictus* the lower three to four or five internodes as well as the top five or six internodes are solid and the central cavity is present only in the rest of the internodes. Variation from species to species in the number of internodes, range in length of internodes and range in wall thickness within a culm of different species of bamboos is presented in Table 6.

The colour of the mature culms also vary in different species being a bright green colour in *Bambusa arundinacea*, a bright yellow colour streaked with green in *Bambusa vulgaris*, and greyish green in *Bambusa polymorpha, Bambusa tulda*, and *Dendrocalamus hamiltonii*.

Pattanath (1965) found significant differences from species to species in the nodal plate i.e. the ladder like partitions at the nodes which interrupt the continuity of the central cavity (pl 11 figs 2 to 6). For instance in *Bambusa nutans* the nodal plate is thick, flat at that bottom and concave and cup shaped at the top with a thickenend rim like portion demarcating the hollowness of the plate. From the outside, the nodal ridge is not completely circular but sloping. In *Bambusa polymorpha* the nodal ridge is double and more or less circular and not sloping. Nodal plate is slightly convex below and concave above. No prominent thickened demarcating rim like portion is

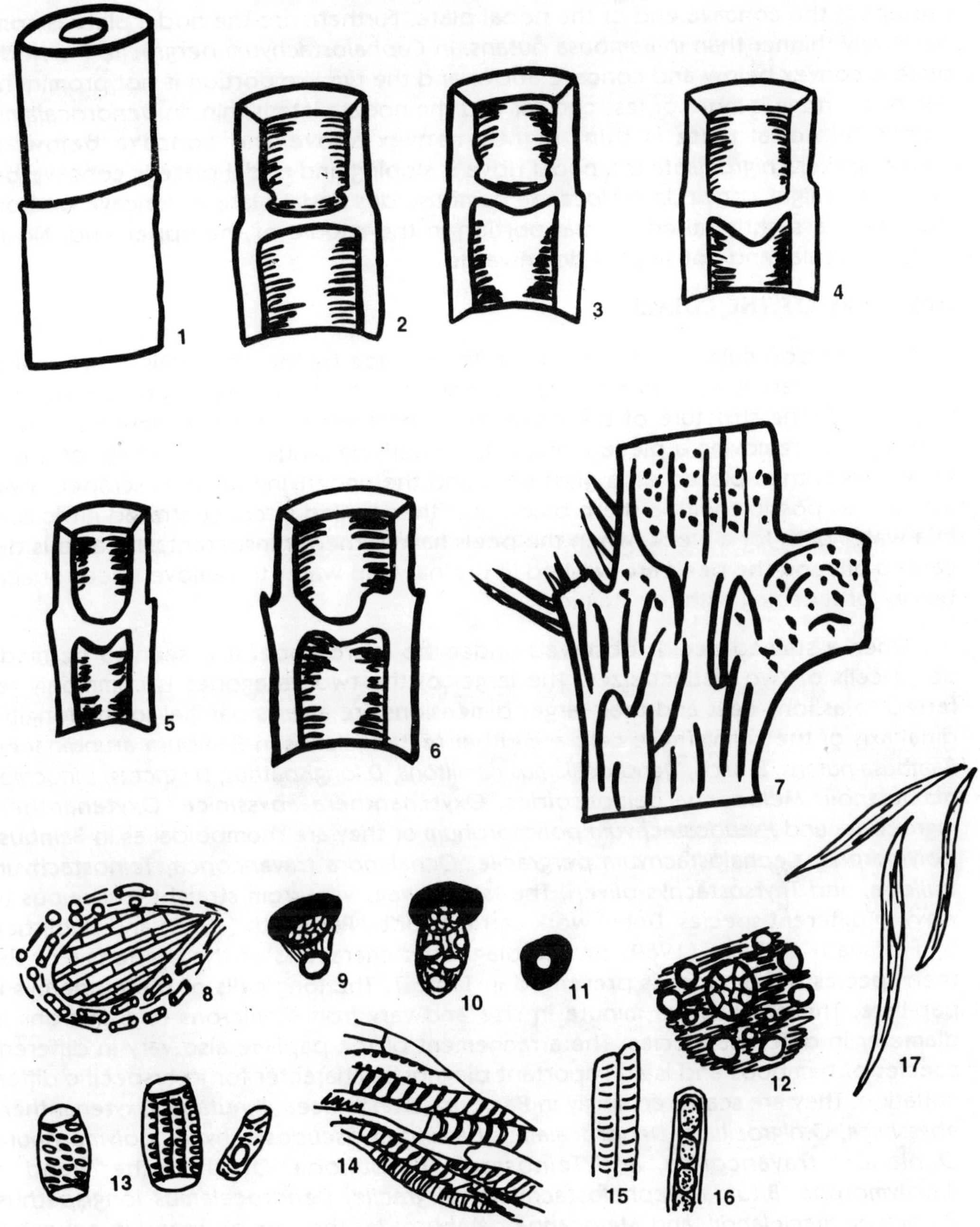

Pl1 11. Nodal structure in bamboos. 1. Nodal portion (exterior view) in *Thyrsostachys oliveri.* 2 to 6. Nodal plates 2. Nodal plate in *B. nutans* 3. in *B. polymorpha* 4. in *Oxytenanthera nigrociliata* 5. in *Melocanna bambusoides* 6. in *Thyrsostachys oliveri* 7. Radial section of Node. 8 to 12 Vascular bundles. 13. Metaxylem. 14. Trace. 15. Protoxylem. 16 Tylosis. 17. Fibres.

present at the concave end of the nodal plate. Furthermore the nodal plate is comparatively thinner than in *Bambusa nutans*. In *Cephalostachyum pergracile* the nodal plate is convex below and concave above and the rim like portion is not prominent. The nodal ridge is more or less circular and the nodal plate is thin. In *Dendrocalamus hamiltonii* nodal plate is thin, slightly convex above and concave below. In *Oxytenanthera nigrociliata* the nodal ridge is sloping and nodal plate is concave below and straight above. In *Melocanna bambusoides* nodal plate is concave on both sides with a slightly raised central portion in the middle of the upper end. Nodal ridge is circular and not sloping downwards.

ANATOMY OF THE CULMS

The bamboo culms are bounded on the outside by the epidermis which makes them very different from all other tree forms of both monocots and dicots (except Sugarcane). The structure of the epidermis is best seen in surface view from peels prepared by removing a piece along with underlying tissue with the help of a pen knife. This is then placed on a glass slide and the underlying tissue is scraped away as much as possible with a sharp blade. It is then boiled in concentrated nitric acid in a water bath for a week, when the peels have turned transparent, the acid is decanted off, and the peels are washed in ordinary tap water to remove traces of acid before proceeding with the staining.

When a stained peel is observed under the microscope. It is seen to be made up of cells of two distinct sizes. The larger of the two categories is commonly referred to as long cells and their larger dimensions are always parallel to the longitudinal axis of the culm. These cells are either rectangular as in *Bambusa arundinacea, Bambusa nutans, B.tulda, Dendrocalamus hamiltonii, D.longispathus, D.strictus, Dinochloa maclelandii, Meloconna bambusoides, Oxytenanthera abyssinica, Oxytenanthera* nigrociliata and *Pseudostachyum polymorphum* or they are rhomboidal as in *Bambusa polymorpha, Cephalostachyum pergracile, Ochalandra travancorica, Teinostachyum dullooa, and Thyrsostachys oliveri.* The lateral walls vary from straight to sinuous to wavy in different species. Detail work carried out by Pattanath (1965) and published by Pattanath and Rao (1969) on the diagnostic characters of the epidermis of sixteen species of bamboos is presented in Table 7. The long cells of the epidermis is papillate. The papillae are minute in size and vary from 2 microns to 6 microns in diameter in different species. The arrangement of the papillae also vary in different species of bamboos and is an important diagnostic character for interspecific differentiation. They are scattered singly in *Bambusa arundinacea. B.nutans, Oxytenanthera abyssinica, O.nigrociliata, Dendrocalamus hamiltonii, Pseudostachyum polymorphum, Ochlandra travancorica, and Teinostachyum dullooa.* On the other hand in *B.polymorpha, B.tulda, Cephalostachyum pergracile, Dendrocalamus longispathus, Dinochloa maclelandii,* and *Melocanna bambusoides* they are arranged in a number of small groups. Further in Thyrsostachys *oliveri* they are arranged in a single large group in the centre of the long cell, and occasionally in two groups (Pl 12 figs 1 to 4)

Each long cell alternates with a pair of cells of the smaller category which appear like dwarf cells as compared with the long cells. Whereas the long cells vary in

Table - 7. Diagnostic Characters of the Epidermis of Sixteen Species of Bamboos.
(Source : - Pattanath 1965)

S. No.	Name of Species	Long cells	Papillae	Stomata	Microhairs
1.	*Bambusa aundinacea* willd.	Rectangular with sinuous to wavy walls.	Scattered singly.	Guard cells sausage shaped & subsidiary cells dome shaped.	Not observed.
2.	*Bambusa nutans* wall.	Rectangular with wavy walls.	Scattered singly.	Surrounded by papillae.	Not observed.
3.	*Bambusa polymorpha* Munro.	Rhomboidal with straight to sinuous walls.	In small gruoups.	Overarched by papillae	Spicules.
4.	*Bambusa tulda* Roxb.	Rectangular with sinuous walls.	In small groups.	Overarched by papillae.	Fan like bicellular hairs.
5.	*Cephalostachyum pergracile* Munro.	Rhomboidal with straight to sinuous walls.	In small groups	Overarched by papillae.	Diverging hairs.
6.	*Dendrocalamus hamiltonii* Nees & Arn.	Rectangular with sinuous walls.	Scattered singly.	Overarched by papillae	Cylindrical bicellular hairs and bladder like hairs.
7.	*Dendrocalamus longispathus* Kurz.	Rectangular with straight walls.	In small groups.	Overarched by papillae.	Spicules.
8.	*Dendro-calamus strictus,* Nees.	Rectangular with straight sinuous or wavy walls.	Scattered singly.	Guard cells sausage shaped & subsidiary cells dome shaped.	Not observed.
9.	*Gigantochloa macrostachya.*	Rectangular with sinuous walls.	Scattered singly.	Dumbell shaped guard cells subsidiary cells absent.	Not observed.
10.	*Melocanna bambusoides* Trin.	Rectangular with sinuous walls.	In small groups.	Overarched by papillae.	Cylindrical bicellular hairs.

11. *Ochlandra travancorica.* Benth.	Rhomboidal with strainght walls.	Scattered singly.	Over arched by papillae.	Diverging hairs.
12. *Oxytenanthera abyssinica* (A Rich) Munro	Rectangular with sinuous walls.	Scattered singly.	Over arched by papillae.	Spicules.
13. *Oxytenanthera nigrociliata* Munro.	Rectangular with sinuous to wavy walls.	Scattered singly.	Guard cells sausage shaped and subsidiary cells dome shaped.	Not observed.
14. *Pseudostachyum polymorphum.*	Rectangular with sinuous walls.	Scattered singly.	Over arched by papillae.	Not observed.
15. *Teinostachyum dulloa.*	Rhomobiodal with straight walls.	Scattered singly.	Dumbell shaped guard cells and subsidiary cells absent.	Not observed.
16. *Thyrsostachys oliveri.* Gamble.	Rhomboidal with straight walls	In single large group.	Guard cells dumbell shaped and subsidiary cells absent.	Spicules.

length from 14 microns to 155 microns in different species of bamboos, short cells vary only from 3.5 microns to 14 microns. The short cells are transversely elongated and always occur in pairs. The shape of these cells vary not only at different heights of the culm but even within an internode (Pattanath 1965). In some species more than one pair alternate with the long cells. The size and frequency of the short cells are useful in differentiating some species (pl2 figs 1 to 8).

Pairs of short cells are sometimes replaced by a stomata or a hair. The structure of the stomata varies in different species of bamboos. In some they are overarched by papillae as in *Bambusa tulda*, *B.polymorpha*, *Dendrocalamus hamiltonii*, *D.longispathus*, *Melocanna bambusoides* and *Ochlandra travancorica*. In some species viz. *Bambusa arudinacea*, *Dendrocalamus strictus*, and *Oxytenanthera nigrociliata* the stomata consists of a pair of sausage shaped guard cells, flanked on either side by dome shaped subsidiary cells. In *Thyrsostachys oliveri*, *Teinostachyum dulloa* and *Gigantochloa macrostachya* the stómata consists of a pair of dumbell shaped guard cells and subsidiary cells are absent. (Pl 12 fig 5 to 8).

Apart from the stomata, another character of diagnostic importance is the type of hairs present. In the epidermis of the bamboo culms a great many types of hairs are present. The large unicellular hairs with broad bases tapering to a point at the apex and which vary in length from 10 microns to 115 microns are known as macrohairs. These are found in almost all species of bamboos especially in the lower

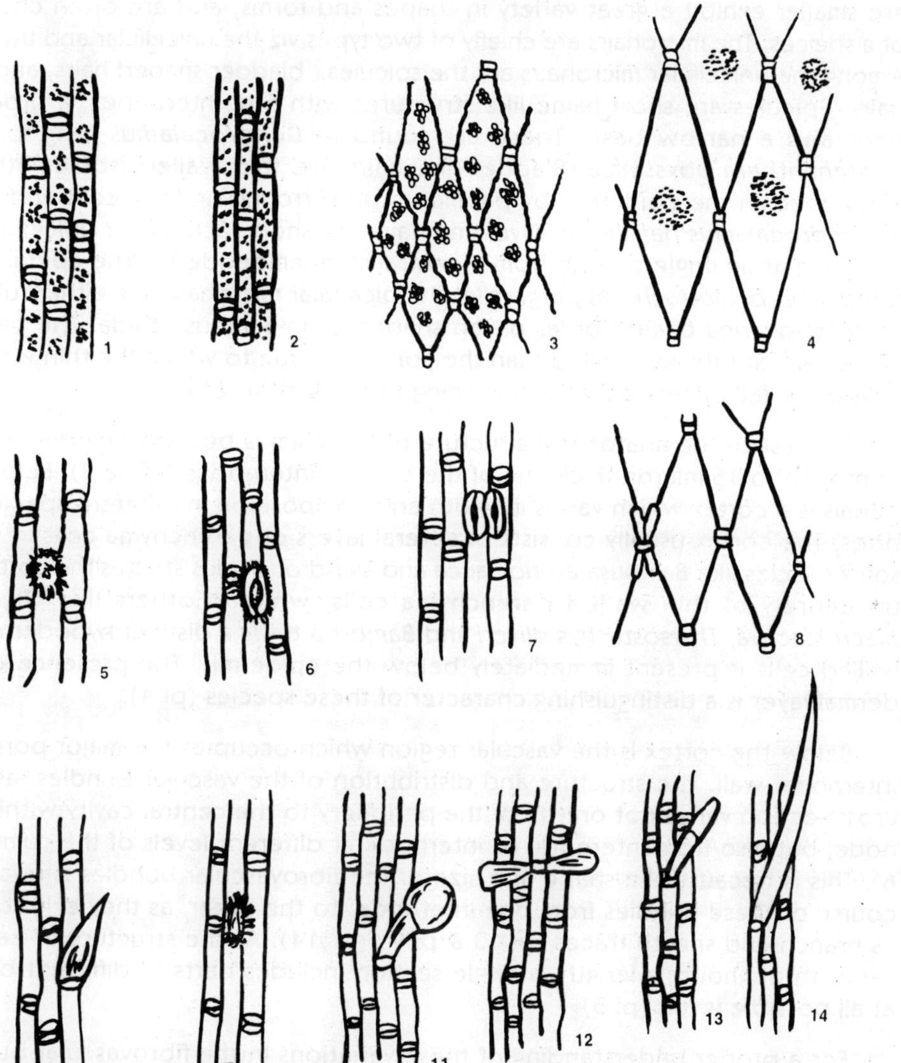

Pl 12. Structural details of culm epidermis of bamboos in surface view. 1. Rectangular long cells with sninuous walls and papillae in groups. 2. Rectangular long cells with sinuous walls and papillae scattered singly. 3. Rhomboidal long cells with sinuous walls and papillae in groups. 4. Rhomboidal long cells with straight walls and papillae in a single large group. 5. Stomata overarched by papillae. 6. Stomata surrounded by papillae. 7. Stomata with sausage shaped guard cells and dome shaped subsidiary cells. 8 Stomata with dumbell shaped gurad cells. 9. Spicule. 10. Bladder shaped microhair. 11. Fan shaped bicellular hair. 12 Diverging hair. 13. Tubular bicellular hair. 14. Macrohair.

internodes. They are not of any importance in identification. The microhairs which are smaller exhibit a great variety in shapes and forms, and are often characteristic of a speices. The microhairs are chiefly of two types viz the unicellular and the bicellular. Among the unicellular microhairs are the spicules, bladder shaped hairs, and diverging hairs. Spicules are short flame like structures with a pointed apex, a broad middle part, and a narrow base. These are found in *Dendrocalamus longispathus* and *Oxytenanthera abyssinica*. Bladder like hairs are thin walled isodiametric bodies embedded in the epidermis by papillae from surrounding long cells and are found in *Dendrocalamus hamiltonii*. Diverging hairs are short rectangular unicellular bodies diverging at an angle of 180° from a central stem embedded in the epidermis. This is found in *Cephalostachyum pergracile*. The bicellular microhairs are either tube shaped as in *Melocanna bambusoides* or fan shaped as in Bambusa tulda. The basal cell of these hairs are thicker walled than the apical cell due to which the thin walled apical cells often fall off while the peel is being prepared (pl12 fig 9.13)

The rest of the anatomical structure of the culm is best observed in a cross section of 10 to 15 micron thickness of the bottom internodes (Table 8). Below the epidermis is a cortex which varies in width and composition in different species of bamboos. The cortex usually consists of several layers of parenchyma cells in section. In some species like *Bambusa arundinacea* and *Dendrocalamus strictus* the cortex is made up entirely of thin walled parenchyma cells, while in others like *Gigantochloa macrostachya, Thyrsostachys oliveri* and *Bambusa tulda* a distinct hypodermis of thick walled cells is present immediately below the epidermis. The presence of a hypodermal layer is a distinguishing character of these species (pl 4).

Below the cortex is the vascular region which occupies the major portion of the internodal wall. The structure and distribution of the vascular bundles, as seen in a cross section varies not only from the periphery to the central cavity within an internode, but also from internode to internode at different levels of the culm (Pl 5 & pl 6). This is because the shape and size of the fibrovascular bundles is related to the course of these bundles from one internode, to the other, as they exit at the nodes as branch and sheath traces (pl 13 & pl 9 fig 7,14). As the structure of each bundle varies throughout its length, a single section includes parts of different bundles cut at all possible levels (pl 5).

For a proper understanding of these variations in the fibrovascular bundles, it is necessary to have a clear conception of the basic structure of the fibrovascular bundle in the Bambuseae. In a complete and fully differentiated bundle (pl 5 fig. 13). There are two metaxylem vassels placed side by side, one protoxylem vessel placed below between the two metaxylem vessels, and one phloem group placed above between the two metaxylem vessels all forming a cross. Each of these conducting elements is supported by a crescent shaped sheath of narrow thick walled fibres. In addition to the sheaths, there are also two patches of fibres (termed as fibre caps for convenience) one associated with the phloem group and the other with the protoxylem group. The various types of vascular bundles found within a culm of any species, and also variations from species to species, are all modifications of the basic structure described above.

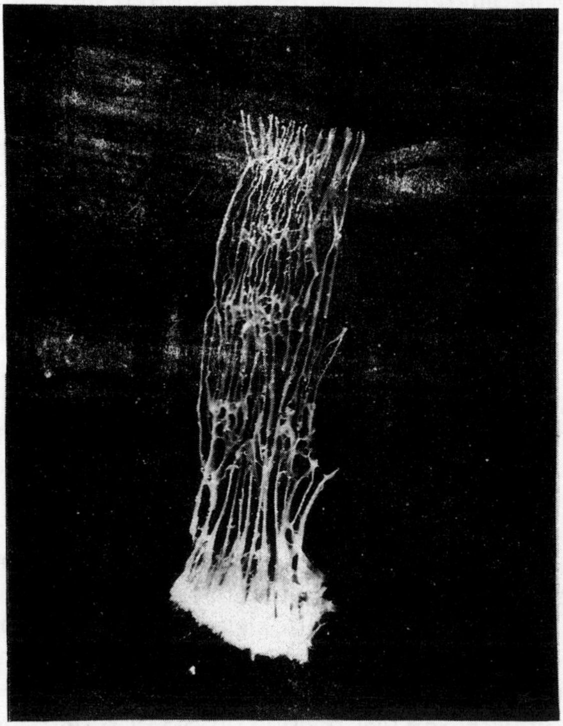

Pl 13. Course of vascular bundles in a young shoot of *Bambusa tulda*.

Based on the form and structure of the bundle, the fibro-vascular region of any internode of any given species of bamboo, can be divided into four intergrading zones viz (i) a narrow peripheral zone, (ii) a transitional zone of varying width, (iii) a wide central zone, and (iv) a narrow inner zone (pl 5). The fibrovascular bundles of corresponding zones in different bamboo speices often show striking differences and are therefore of considerable help in identification (Pl 3a & 3B).

In the peripheral zone, the vascular bundles in some species e.g. *Bambusa tulda* are reduced to single metaxylem vessel surrounded by narrow fibre sheath forming a band of two or three layers. In *Bambusa nutans* the peripheral zone consists of closely spaced vascular bundles some of which are reduced to a single vessel surrounded by a narrow sheath of fibres, and others are small complete bundles consisting of a phloem group, and two metaxylem and a protoxylem. The phloem fibre sheath is circular in shape and much reduced in size. The three xylem vessels are joined together with a single large sheath which is circular to cordate in shapes. In *Dendrocalamus strictus* the vascular bundles in the peripheral zone are widely spaced. Patches of phloem are scattered among small complete bundles, the fibre sheaths of which are again unequal in size. Phloem fibre sheath is much reduced in size and xylem fibre sheath is large and circular to crescent shaped. In *Dendrocalamus hamiltonii* the vascular bundles of the peripheral zone is widely spaced small complete bundles. The phloem fibre sheath is reduced in size and knob like in appearances. The xylem fibre sheath is circular to cordate in shape. Independent patches of fibres circular to cordate in shape are also often scattered among the peripheral vascular bundles. In *Oxytenanthera nigrociliata* the peripheral zone consists of patches of phloem as well as reduced bundles consisting of only a single vessel supported by a crescent shaped sheath of fibres, and small complete bundles surrounded by radially elongated oval sheath of fibres which are lager towards the xylem. In *Oxytenanthera abyssinica* the peripheral vascular region consists of widely spaced bundles of tracheids scattered among small complete bundles with phloem fibre sheath reduced in size, and xylem fibre sheath large and oval in shape. (pl 5 fig 1 to 7).

In the transitional zone, the bundles are larger than in the peripheral zone, but smaller than the central and inner zones. Further, in the vascular bundles of the transitional zone there are only two sheaths of fibres viz one supporting the phloem and the other supporting the three xylem vessels, coalesced into one massive sheath. In addition to the two sheaths, in some species eg *Dendrocalamus hamiltonii*, the phloem fibre cap or the xylem fibre cap, or both fibre caps may be present (pl 5 fig 8 to 12). There is thus a mixture of more than one type of vascular bundles in the transitional zone of some species. Moreover, the shape and size of the fibre sheaths, and frequency of the fibrovascular bundles also vary from species to species. For instance in *Bambusa nutans* and *Dinochloa maclelandii* the bundles are closely spaced with vary little parenchyma in between. On the other hand in *Oxytenanthera abyssinica* they are relatively widely spaced.

In the central zone the vascular bundles are fully differentiated, and each of the four conducting elements has its own crescent shaped sheath of fibres, as well as one or two fibre caps. The fibre caps are normally separated from the sheaths by parenchyma. In some species however one or more fibre caps may be joined to the respective sheaths. In *Bambusa polymorpha* for instance the phloem fibre cap is joined with the phloem fibre sheath. In Cephalostachyum pergracile there appears to be four fibre caps each of which is coalesced with its respective sheath i.e. phloem fibre cap with phloem fibre sheath, protoxylem fibre cap with protoxylem fibre sheath, and the two metaxylem fibre caps with the metaxylem fibre sheaths. In *Dinochloa maclelandii* there is only the xylem fibre cap present and it is separated from the protoxylem fibre sheath. In *Thyrsostachys oliveri* and *Oxytenanthera nigrociliata* two fibre caps are present viz the xylem fibre cap and the phloem fibre cap. Both caps are separated from their respective sheaths (Pl 5 figs 13 to 17). In species with two fibre caps to each bundle, the central zone often has a mixture of bundles - some with two fibre caps and some with only one xylem fibre cap present. Furhtermore in these species, the bundles of the central region in the higher internodes have only the xylem fibre cap. The phloem fibre cap is suppressed and dispappears in the higher internodes (Pl 6.)

In the inner zone both fibre caps disappear and the bundles consist of the four conducting elements each supported by a crescent shaped sheath of fibres (Pl 5 fig 21). The bundles however are large in size. In between the bundles of the central and inner region there are in some species bundles resembling those of the transitional zone (Pl 5 fig 20). In the bundles of the peripheral, transitional and central zones the phloem group is always oriented towards the epidermal side, and the protoxylem towards the central cavity (Pl 5 to 1 to 17). But in the inner zone the phloem group is often turned towards the central cavity (Pl 5 fig 19 to 20). The phloem group of each bundle consists of several sieve tubes of varying sizes and companion cells. The sieve tubes towards the epidermal side have a smaller cross sectional diameter than those towards the inner side. In between the sieve tubes are the companion cells. Each sieve tube seem to be containing several companion cells. Opposite to the phloem group is the protoxylem which is sometimes replaced by a lacunae. Metaxylem vessels are mostly barrel shaped with elongated pits arranged irregularly on the walls. The perforation plates are simple and terminal.

The structure of the vascular bundles is altered as they enter the nodes. The fibre cap of the phloem pole of the bundle is greatly enlarged while the fibre cap of the xylem pole has nearly disappeared (Pl 11 fig 9 to 10). The nodal bundles lack (i) the protoxylem lacunae and (ii) the fibre sheaths around the vessels. There is also a general increase in the number of protoxylem elements some of which are filled with tylosis (Pl 11 fig 15 to 16). The metaxylem vessels show a tendency to increase in number and decrease in size. Most of them are small and barrel shaped with elongated pits arranged irregularly on their walls. The perforation plate though simple are not always terminal in position. Sometimes the perforation plates are found on the lateral walls (Pl 11 fig 12 to 13). The phloem shows a quantitative increase over that found in the internodes. The fibres which surround the bundles are also greatly reduced in length and is often less than half the length of fibres of the internodes (Pl 11 Fig 17). The traces which enter the node from the culm sheath deviate from the normal structure in their arrangement of xylem and phloem, which instead of being distinctly collateral, approaches the amphivasal type, and instead of the large pitted metaxylem vessels, there is an abudndance of narrow elements with their secondary

Table 8. Diagnostic Characters of the Internal Structure of Culm of some Species of Bamboos in the Bottom Internodes (Source:Pattanath 1965)

S. No.	Name of Species	Cortex	Peripheral f.v.b.	Transitional f.v.b.	Central f.v.b.
1.	*Bambusa arundinacea*	Cortex of thin walled closely packed parenchyma cells and intercellular spaces absent. Hypodermis absent.	Small complete bundles with unequal phloem and xylem sheaths mixed with a few reduced bundles of only metaxylem.	Phloem fibre sheath small and circular, xylem fibre sheath large & vary in shape from cordate to lunate.	Both fibre caps present and usually unequal in size. Phloem fibre cap larger & vary in shape from circular to reniform. Xylem fibre cap small in size and circular in shape. phloem and metaxylem sheath nearly same size. Protoxylem sheath much reduced.
2.	*Bambusa nutans*	Cortex made up of thin walled closely packed parenchyma. Intercellular spaces absent. Hypodermis absent.	A few reduced bundles scattered among closely spaced small complete bundles with phloem and xylem sheaths of unequal size.	Phloem fibre sheath small and circular in shape. Xylem fibre sheath large and circular to lunate in shape.	Both fibre caps present in a few bundles, but mostly phloem fibre cap absent. When present both caps of equal size and circular in shape - Phloem fibre sheath larger than the metaxylem fibre sheaths, Protoxylem sheath reduced, and small in size.
3.	*Bambusa polymorpha*	Cortex made of thin walled parenchyma cells. Intercellular spaces absent. Hypodermis absent.	A few reduced bundles consisting of single metaxylem vessel scattered among complete bundles with phloem fibre sheath much	Phloem fibre sheath circular to knob like and xylem fibre sheath lunate in shape.	Both caps present and nearly of the same size. Phloem fibre cap attached to phloem fibre sheath. Xylem fibre cap separated from xylem fibre sheath.

			reduced and xylem sheath large and circular to cordate in shape.		
4.	*Bambusa tulda*	Hypodermis of thick walled elongated cells arranged in groups. Cortex of thin walled closely packed cells. Intercellular. spaces absent.	A band of reduced vascular bundles consisting of a single sometimes two metaxylem vessels surrounded by fibres	Phloem fibre sheath small and more or less circular in shape. Xylem fibre sheath large and vary in shape from circular to cordate.	Some bundles with both fibre caps. Xylem fibre cap usually larger than phloem fibre cap. In some bundles phloem fibre cap absent. Phloem and xylem sheaths nearly of the same size.
5.	*Cephalostachyum pergracile.*	Cortex with closely packed thin walled parenchyma cells. Intercellular spaces absent. Hypodermis absent.	Reduced bundles scattered along with complete bundles with phloem fibre sheath greatly reduced and xylem fibre sheath oval to cordate in shape and much larger than the phloem sheath.	Phloem fibre sheath circular and knob like. Xylem fibre sheath large and circular to cordate in shape.	Both fibre caps present and joined to their respective sheaths. In addition to these there are also two metaxylem caps joined to the metaxylem sheaths. All caps crescent to lunate in shape.
6.	*Dendrocalamus hamiltonii.*	Cortex made up of thin walled parenchyma. Intercellular spaces absent Hypodermis absent.	Widely spaced complete small bundles with phloem fibre sheath knob like xylem fibre sheath circular to cordate in shape and patches of fibres scattered among the bundles.	Two types of bundles present. Bundles with only the sheaths present, and bundles with sheaths as well as one or two fibre caps also. Phloem fibre sheath circular in shape and smaller than xylem fibre sheath which is more or less crescent shaped.	Both fibre caps present and mostly of nearly the same size occasionally unequal. Shape of caps vary from circular to cordate to reniform. Phloem and metaxylem fibre sheaths of nearly the same width. Protoxylem fibre sheath very reduced in size.

7.	*Dendrocalamus longispathus.*	Cortex made up of thick walled parenchyma with intercellular spaces. Hypod--ermis absent.	Widely spaced complete bundles with phloem sheath reduced and xylem sheath large and more or less circular in shape	Phloem fibre sheath circular and knob like and xylem fibre sheath large and lunate to cordate in shape.	Both fibre caps present. Phloem fibre cap large cordate tc reniform in shape. Xylem fibre cap small and vary from circualr to cordate and reniform. All fibre sheaths moderatly broad.
8.	*Dendrocalamus strictus*	Cortex made up of thin walled parenchyma. Intercellular spaces absent Hypodermis absent.	Widely spaced small complete bundles with phloem sheath much reduced and xylem fibre sheath large circular to crescent shaped. Independent pat--ches of phloem scattered among the bundles.	Xylem fibre sheath only slightly larger than phloem fibre sheath. Both sheaths circular in shape. Xylem sheath sometimes radially elongated. Independent patches of fibres scattered among the bundles.	Two types of bundles present. Mostly bundles with only xylem fibre cap. Some bundles have both fibre caps but phloem fibre cap is much reduced - xylem fibre cap vary in shape from reniform to cordate. In some bundles the phloem fibre cap is joined to the phloem fibre sheath. Metaxylem and phloem fibre sheaths moderately wide. Protoxylem fibre sheath much reduced.
9.	*Gigantochloa macrostachya.*	Cortex of thick walled parenchyma cells with inter cellular spaces. Hypodermis absent.	Complete vascular bundles with crescent shaped phloem sheaths and oval xylem sheaths.	Oval shaped phloem sheaths smaller than the large oval shaped xylem sheaths.	Only one xylem fibre cap present and separate from the xylem sheath. Phloem fibre cap absent. Phloem fibre sheath and metaxylem fibre sheath of equal size.
10.	*Melocanna bambusoides*	Cortex made up of thick walled paren--chyma. Intercellular spaces present. Hypodermis absent.	Patches of fibres scattered among small complete bundles with phloem fibre sheath greatly reduced. Xylem fibre sheath vary	Phloem fibre sheath small & circular in shape.Xylem fibre sheath large and circular to cordate in shape.	Three types of bundles present (i) In some, phloem fibre cap absent and xylem fibre cap present; (ii) In some bundles xylem cap joined to the xylem fibre sheath, (iii) In some bundles the xylem fibre cap is

		in shape from oval to circular to cordate		separated from the xylem fibre sheath. In all bundles all four fibre sheaths are large and crescent shaped.
11. *Oxytenanthera abyssinica.*	Cortex made up of thin walled parenchyma -cells. Intercellular spaces absent. Hypodermis absent.	Small complete widely spaced bundles scattered among bundles of trachieds.	Phloem firbre sheath small and circular. Xylem fibre sheath large and circular to oval in shape.	Both fibre caps present. Phloem fibre cap joined to phloem firbe sheath. Xylem fibre cap separated from the sheath. Metaxylem and protoxylem fibre sheaths very narrow. Phloem fibre sheath moderately wide.
12. *Oxytenanthera nigrociliata.*	Hypodermis absent. Cortex of thin walled parenchyma. Inter--cellular spaces absent.	Small compete bundles with radially elongated xylem sheaths and reduced phloem sheaths scattered among reduced bundles consisting of a single metaxylem vessel supported by a narrow sheath. Independent patches of phloem also present.	Phloem fibre sheath knob like and circular, xylem fibre sheath crescent shaped. Xylem fibre cap joined to the sheath and oval in shape. Phloem fibre cap absent.	Both fibre caps present and unequal in size. Phloem fibre cap cordate to reniform in shape. Xylem fibre cap circular to cordate. Phloem and metaxylem fibre sheath moderately wide protoxylem fibre sheaths sometimes reduced.
13. *Thyrsostachys oliveri.*	Hypodermis present, Made up of thick walled rounded cells, and cortex of thick walled parenchyma cells. Intercellular spaces present.	Widely spaced complete small bundles with large xylem sheaths lunate to cordate in shape. Phloem fibre sheath greatly reduced and nearly absent.	Phloem fibre sheath small and conical, xylem fibre sheath conical and elongated. Some bundles with both fibre caps present. Phloem fibre caps separated from the sheath. Xylem fibre cap joined to the sheath	Both caps present. Phloem fibre cap large and cordate xylem fibre cap small and circular in shape. Fibre sheaths moderately wide.

thickenings in the form of very close spirals (Pl 11 fig 14). The epidermis in the nodal region is very thick walled and bears long unicellular hairs. The parenchyma is thin walled as compared with those of the internodes and is isodiametric in shape. In the internode the parenchyma is more or less rectangular in shape as observed in radial and tangentical sections, and consists of some longer cells than others giving the impression of long cells and short cells as in the epidermis. But unlike in the epidermis they do not alternate with each other (Pl 8 fig 2 to 6). The parechyma also are more or less thick walled as compared to the parenchyma of the nodes. The nodal plate is bordered by a band of thick walled sclerenchymatous cells above and below.

In the internode the fibrovascular bundles follow a longitudinal course and run approximately parallel to each other. But when they reach the node, some of them give out transverse branches, and some bend abruptly and move towards the periphery. At the level of the nodal plate the vertical and horizontal bundles anastomose forming a plexus. The bundles are more crowded towards the periphery, and some of the peripheral bundles pass out into the culm sheath. In the mature culms, where the culm sheath has fallen off, the bundles passing out into the sheath, is cut off by two to three layers of thick walled cells, so that in the mature culms, the bundles passing into the culm sheath end abruptly. The traces which enter the node from the culm sheath penetrate horizontally to the nodal plate, so that in a cross section they appear like rays between the vertical bundles. The number of traces that enter the node from the culm sheath varies at different levels of the culm. The number of bundles passing from the culm to the culm sheath decreases in number in successive internodes starting from the bottom of the culm upwards.

Pattanath (1965) studied the course of vascular bundles within a culm of *Bambusa tulda* from a small shoot collected from the Demonstration Area of the New Forest Estate in Dehradun. The whole shoot was bleached in Hydrogen peroxide and cleared in lactic acid. It was then possible to tease out the vascular bundles with the help of a dissection needle (Pl. 13). It was found that the rhizome portion itself had 106 vascular bundles but in the first internode only 82 were visible and it kept decreasing to 77 in the second internode, 69 in the third, 58 in the fourth, 52 in the fifth and so on. It was also observed that the fibrovascular bundles were widely distributed towards the margin of the cross section and became close together at about the middle region. Bundles of the internode run parallel to each other and anastomose at the nodes and is redistributed in the next internode. A large number of horizontal branches enter the nodal plate from the periphery of the node and run horizontally into the nodal plate sometimes anastomosing with longitudinal bundles. Traces from the culm sheath enter the nodal plate and run horizontally towards the centre in between the longitudinal bundles of the internode giving a ray like appearance.

In all bamboo species three types of fibres are distinguished viz. (i) sheath fibres of the fibrovascular bundles of the internodes (ii) cap fibres of the fibrovascular bundles of the internodes (iii) fibres of the fibrovascular bundles of the nodes. The sheath fibres of the fibrovascular bundles of the internodes are long narrow thick walled with their tapering end portions drawn out into a fine point. The cap fibres of the

Table 9. Fibre dimensions of twelve species of bamboos (at 4 different heights of culms)

Name of species	*Average fibre length at diff. heights of the culm in microns*				*Average fibre width at diff. heights of the culm in microns*				*Average fibre wall thickness at different heights of the culm in micron*				*Slenderness ratio of fibres at diff. heights of the culm (L/D)*			
	Bott.	¼ht.	Mid. ht.	¾ht.	Bott.	¼ht.	Mid. ht.	¾ht.	Bott.	¼ht.	Mid. ht.	¾ht.	Bott.	¼ht.	Mid. ht.	¾ht
Bambusa arundinacea	1946	1994	2286	1974	21.7	19.2	18.9	20.8	6.4	6.3	6.0	5.9	89	103	120	94
B.nutans	1902	2303	1737	1420	21.7	22.4	18.2	20.5	6.9	6.8	5.4	6.9	87	102	95	68
B.polymorpha	1805	2100	1627	2220	20.4	20.7	20.7	20.3	5.6	7.3	8.2	7.4	88	101	78	108
B.tulda	2033	1953	2095	1638	20.4	20.5	18.9	18.6	7.6	5.6	6.4	6.2	99	94	110	87
Cephalostachyum pergracile	2156	1862	1906	1957	19.1	17.0	22.0	20.3	6.4	5.7	7.3	6.0	112	109	86	96
Dendrocalamus hamiltonii	2108	2572	2291	1463	17.5	20.2	17.7	17.2	4.5	6.3	5.4	6.1	120	124	129	85
D.longispathus	1862	2453	2340	1607	21.9	22.1	22.7	18.9	5.5	7.6	7.9	7.2	84	110	102	84
D.strictus	2114	2190	1906	1912	25.4	23.8	23.5	24.1	8.7	8.6	8.9	8.1	83	92	81	79
Melocanna bambusoides	1952	2158	1782	2043	22.0	21.1	21.7	19.6	7.1	7.1	8.8	7.3	87	102	82	103
Oxytenanthera abyssinica	1858	2075	1901	1866	22.0	20.0	17.0	20.0	7.0	5.1	5.3	7.6	84	103	111	93
O. nigrociliata	2543	2332	1943	1823	18.1	21.7	22.4	19.1	4.7	6.8	8.8	7.5	140	107	86	96
Thyrsostachys oliveri	2016	1921	1891	1870	21.7	19.1	20.0	15.1	7.8	7.1	7.9	5.6	92	100	94	123

TABLE -9

(Source : Pattanath 1965)

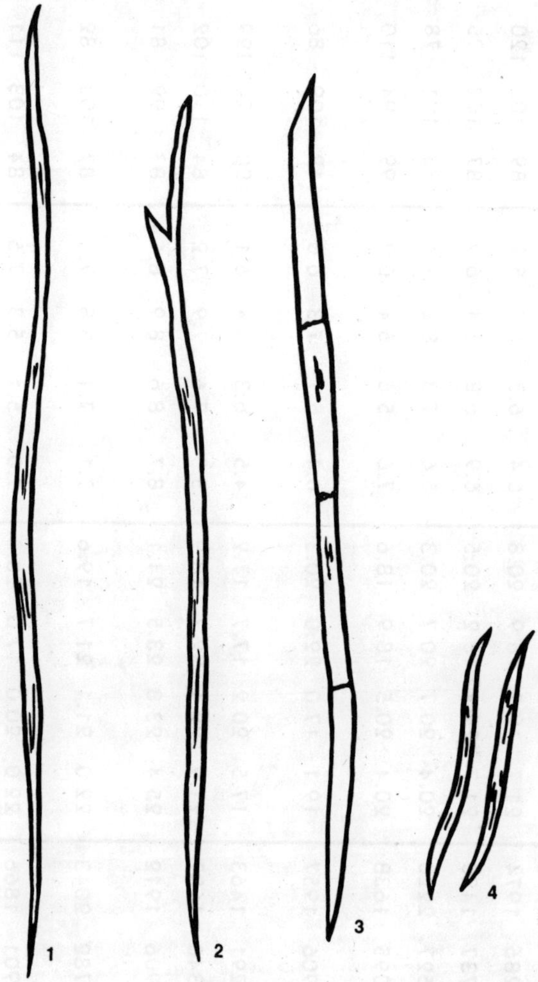

Pl 14. Structure of bamboo fibres. 1. Sheath fibres of the internodes of culms. 2&3. Cap fibres of internodes of culms. 4. Nodal fibres.

fibrovascular bundles on the other hand are relatively broader than the sheath fibres, thinner walled with their tapering ends blunt with rounded tips, oblique tips, and sometimes bifurcated tips. The pits on the fibres are arranged in radial rows along the longitudinal axis. The borders of these pits are very reduced in size and are barely visible even at high magnification (Pl 14 figs 1 to 3). The fibre dimension of the internodes vary considerably from internode to internode within a culm and also from species to species. Variations of fibre dimensions of internodes from four different levels of the culm of twelve species of bamboos are presented in Table 9. The fibres of the fibrovascular bundles of the nodal portion are less than half the length of those of the internodal portion. They are also narrower and thicker walled and resemble the sheath fibres of the fibrovascular bundles of the internode (Pl 14 Fig 4)

There is a nonvascular region adjacent to the central cavity in the internodes which in some species show striking differences and are therefore useful in differentiating them. In *Dendrocalamus longispathus* it consists of a continuous band of sclerenchyma, while in *Cephalostachyum pergracile* the sclerenchymatous cells occur as irregular patches tending to form an interrupted band. In majority of species the lining of the central cavity is parenchymatous (Pl 7).

CULM SHEATH

The culms of all species of bamboos when young are enclosed in large generally coriaceous sheaths often hairy outside. The sheaths arise at the nodes and as a rule terminate in a blade of varying shapes and structures in the different species. The sheaths are strictly alternate and there is an angle of 180° between the axillary buds of two successive sheaths. The general apperance, length, texture, shape and colour of the sheaths vary between species and may be used as distinguishing characters for different species. A sheath is made up of two parts viz. the sheath which envelops the culm, and the blade of varying shapes and structures. Between the sheath and the blade is a structure known as the ligule, and the blade in some species has a structure known as the auricle. (Pl 9 Fig 4 & 5)

MORPHOLOGY OF THE CULM SHEATH

Chatterjee and Raizada (1963) worked on the morphology of culm sheaths of 22 species of bamboos growing in the Forest Research Insititue Campus at New Forest Dehradun. They found that though the culm sheaths at first appear distinctive for each species there is considerable variation within a culm in the size of the culm sheaths as well as in the shape of the blade. However it was possible to distinguish some species from others by the colossal size of the sheaths of *Dendrocalamus ginganteus, D.calostachyus* and *D.brandisii* as compared to the others. Further, the cylindrical shape of the sheaths are characteristic of *Dendrocalamus longispathus* and *D.membranaceous,* while the oblique asymmetrical sheaths are characteristic of *Bambusa nutans* and *B.tulda,* and the conical shape of the sheaths are characteristic of *Bambusa pallida, Dendrocalamus strictus* and *D.hamiltonii.* In some cases as in *Bambusa tulda, B.nutans* and *B.burmanica* the sheaths look very much alike at first. However the auricles in *B. tulda* are quite distinct from the blade and very much laterally placed on top of the sheath. In *B.nutans* on the other hand, the auricles are more or less con-

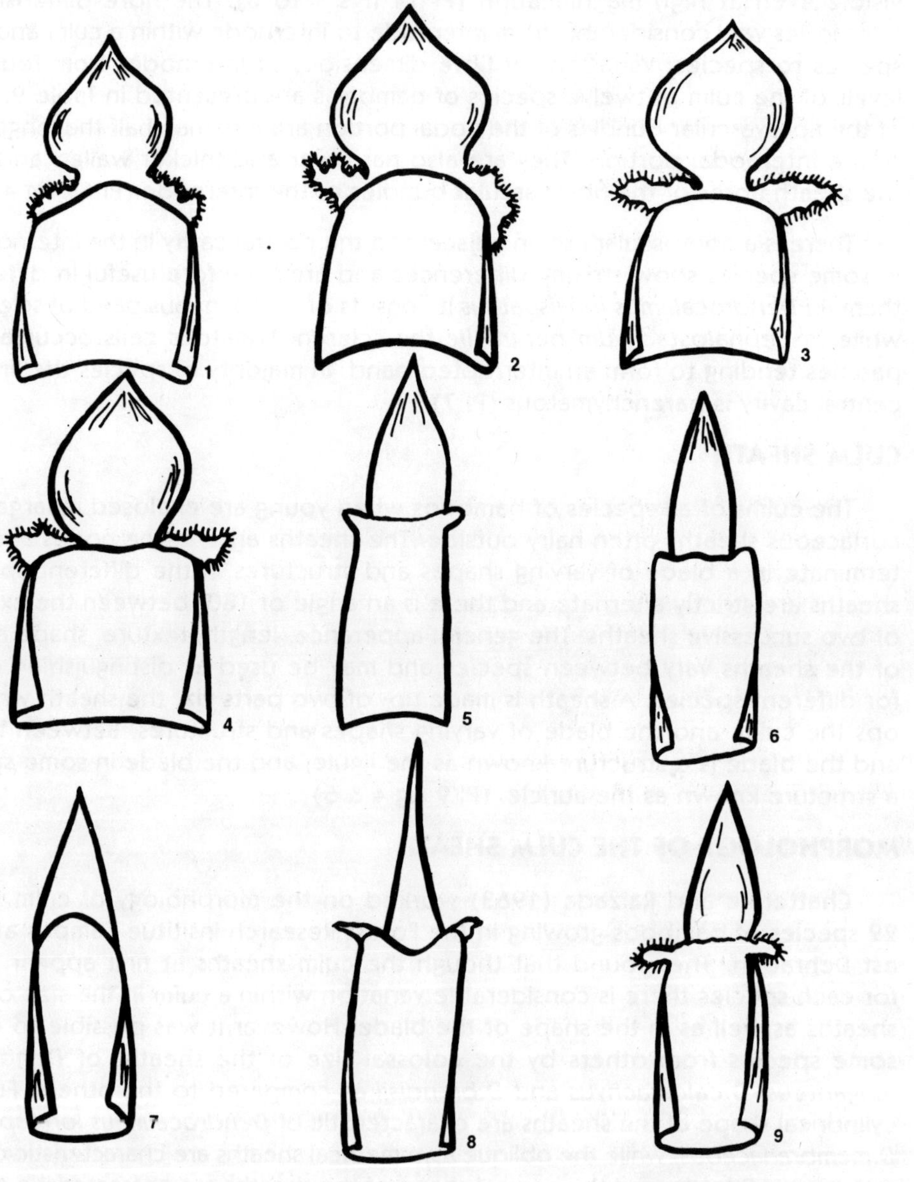

Pl 15. Culm sheath in different bamboo species. 1. *Bambusa arundinacea* 2. *B. tulda* 3. *B. polymorpha* 4. *Cephalostachyum pergracile* 5. *Dendrocalamus hamiltonii* 6. *D. longispathus* 7. *D. strictus* 8. *Melocanna bambusoides*. 9. *Oxytenanthera nigrociliata*.

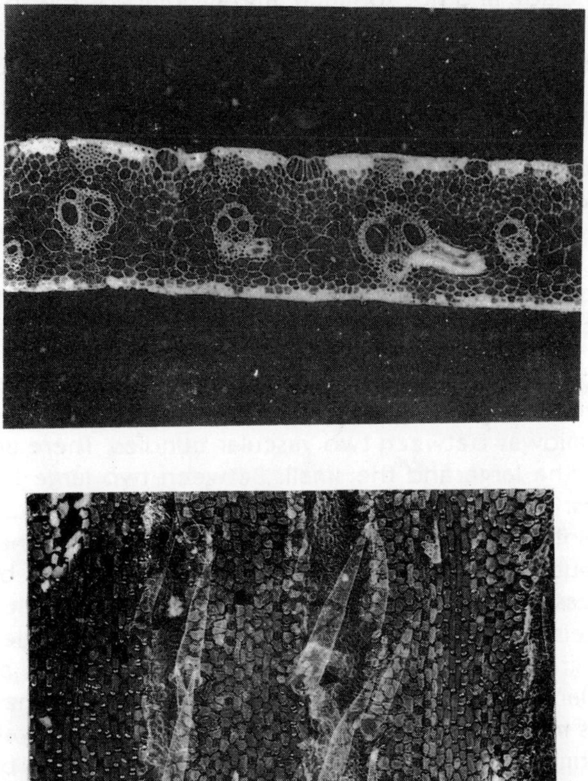

Pl 16. Anatomy of culm sheaths. 1. Cross section of culm sheath of *B. tulda* (Magnification 43X) Surface view of outer epidermis of culm sheath of *B. tulda* (Magnification 60X) 3. Epidermal cells and stomata enlarged (Magnification 400X).

tinuous with the blade and not so much laterally placed. The sheaths of *B.tulda* and *B.burmanica* are however difficult to differentiate. In *Melocanna bambusoides* the blade is subulate and pointed with very prominent auricles (Pl 15).

ANATOMY OF THE CULM SHEATH

Some preliminary observations carried out on the anatomy of the culm sheath of *Bambusa tulda* shows that the epidermis in surface view consists of squarish long cells alternating with short cells and are at intervals replaced with rows of long cells alternating with stomata, which consists of sausage shaped guard cells and dome shaped subsidiary cells. At regular intervals are also found rows of long narrow macro hairs with pointed apex. (Pl 16 Fig 1& 2)

In a cross section of the culm sheath the outer epidermis appears fairly thick walled and is interrupted at intervals by groups of bulliform cells. These groups of bulliform cells are placed midway between two vascular bundles. There are two types of vascular bundles — the large and the small. Between two large bundles are three to four small bundles. The large vascular bundles consist of two metaxylem vessels placed side by side. a protoxylem vessel placed between and below the metaxylem vessels, and facing the outer epidermis, and a phloem group placed between and above the metaxylem vessels and facing the inner epidermis. The smaller bundles consist of 1 to 4 small vessels and a phloem group surrounded by a single layer of small thick walled fibres. A group of similar thick walled fibres also form a patch immediately below the vascular bundle just next to the outer epidermis. The inner epidermis on the other hand is made up of an uninterrupted layer of thick walled cells. The outer and inner epidermis is connected by thin walled isodiametric parenchyma cells (Pl 16 fig 3).

LEAF

MORPHOLOGY OF LEAF

Leaves in bamboos are distichous and consists of a tubular sheath split to the base and a linear, oblong or lanceolate blade with a midrib and numerous longitudinal veins. The veins are of two types (i) stout and (ii) thin. The thin veins range from 2 to 8 between two stout veins. The blade is joined to the sheath by a short petiole. All bamboo leaves have transverse veins which connect two longitudinal nerves. These transverse veins may be straight, oblique, or with a bend in the middle (pl 17 fig 1&2). Pieces of dry leaves boiled in water and examined in glycerine, makes the transverse veins very clear. The leaves being distichous, an inner and an outer edge may be distinguished. The inner edge is generally fringed with short thick walled sharply pointed hairs, while the outer edge has long thick walled hairs eg. *Thyrosostachys oliveri*. In some species both edges are fringed with sharp pointed thick walled hairs, and in others on the underside of the leaf, the epidermis has numerous solid protruberances or hairs which are short and thick walled or long and soft (Pl.17 fig 1).

Brandis (1921) found that identification of many species of bamboos, when only leaves are available, may be facilitated by counting the number of longitudinal nerves on a quarter inch of the leaf. He found that in most genera the number varies from

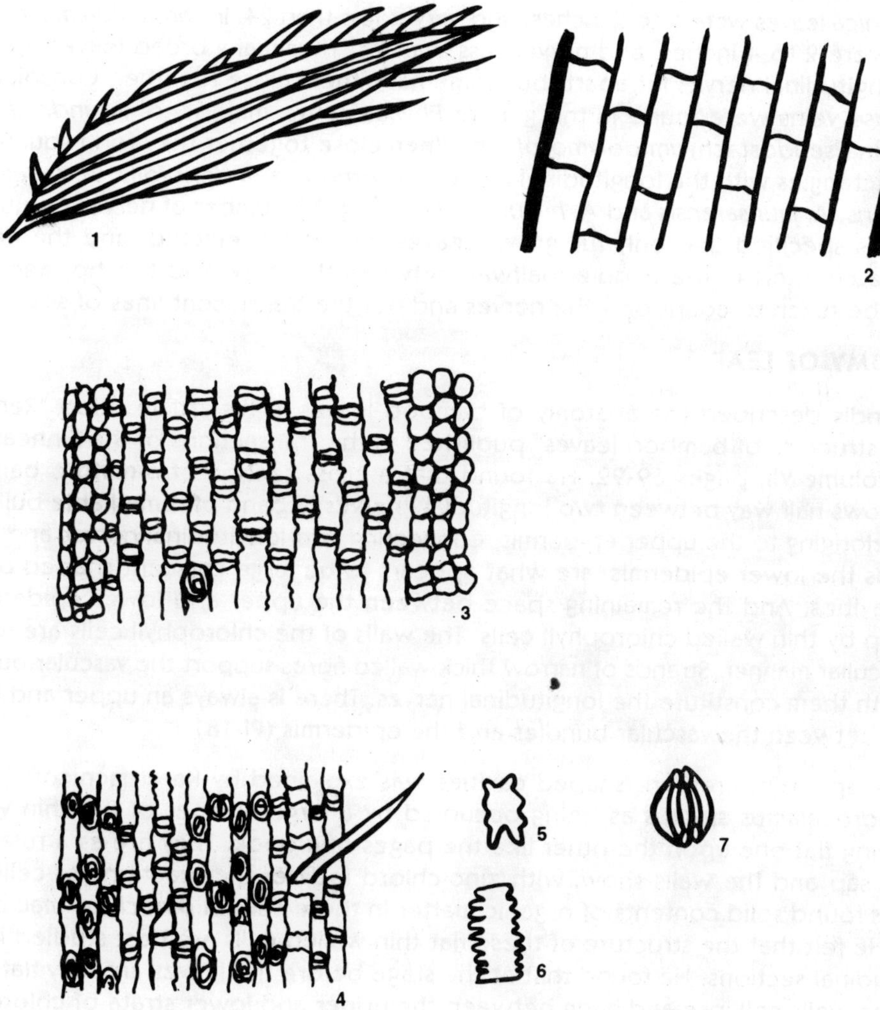

Pl 17. Morphology and anatomy of bamboo leaves. 1. Leaf of *Thyrsostachys oliveri* 2. Stout veins, slender veins and transverse veins. 3. Upper epidermis of leaf 4. Lower epidermis of leaf 5. Stomata bearing epidermal cell. 6. Long cell alternating with short cells. 7. Stomata.

20 to 50. *Dendrocalamus* and *Melocalamus* were found to have 20 to 30. In *Arundinaria walkeriana* the leaf size was observed to be 1 to $^2/_3$ inch wide and longitudinal nerves were less than 24. In *A. hookeriana* leaf width was ½to 1½inch and longitudinal nerves less than 24. In *Bambusa khasiana* leaves were 1 to 1½inch wide and longitudinal nerves less than 24. In *B. griffithiana* leaves were 3 inches, but longitudinal nerves were less than 24. In *Pseudostachyum polymorphum* and *Dinochloa andamanica* leaves were 1 to 2 inches, and nerves less than 24. In *Dinochloa maclelandii* leaves were 2 to 4 inches, and nerves less than 24. Thus many broad leaved species have longitudinal nerves far apart, but some have them close together. Conspicuous transverse veins were found in the genera *Phyllostachys,* majority of *Arundinaria,* as well as in *Pseudostachyum polymorphum.* When close together they form squares or short rectangles with the longitudinal nerves eg. *Arundinaria densifolia, A. racemosa, A. elegans, A. jaunsarensis* and *A. hirsuta.* In counting the number of nerves on quarter inch, it is specified that only full grown leaves should be selected, and the nerves must be counted in the middle, halfway between the base and the tip, and care should be taken to count only the nerves and not the translucent lines of silica cells.

ANATOMY OF LEAF

Brandis described the anatomy of bamboo leaves in his article titled "Remarks on the structure of bamboo leaves" published in the Transactions of the Linnean Society, volume VII, pages 69-92. He found that a cross section of a mature bamboo leaf, shows half way between two longitudianl nerves, a band of remarkable bulliform cells belonging to the upper epidermis. Connecting two longitudinal nerves and more towards the lower epidermis, are what appears to be large crescent shaped or oblong cavities. And the remaining space between the upper and lower epidermis is filled up by thin walled chlorophyll cells. The walls of the chlorophyll cells are folded in a peculiar manner. Strands of narrow thick walled fibres support the vascular bundles and with them constitute the longitudinal nerves. There is always an upper and lower strand between the vascular bundles and the epidermis (Pl 18).

The apparent crescent shaped cavities was explained by Karelstschicoff (1868) in *Dendrocalamus strictus* as being occupied by a large number of flat thin walled cells lying flat one upon the other like the pages of a book. They are as a rule filled with a sap and the walls show, with zinc chloro iodide, the reactions of cellulose. Brandis found solid contents of organic matter in these cells in *Dinochloa maclelandii* Kurz. He felt that the structure of these flat thin walled cells are best studied in thin longitudinal sections. He found that at the stage before the leaves are fully flattened out, the walls collapse and hang between the upper and lower strata of chlorophyll parenchyma resembling fine empty bags. This he found in *Arundinaria japonica, Phyllostachys bambusoides* and *Dinochloa andamanica.*

A general character of the bamboo leaf is that the lower epidermis is uneven with protruberances and the upper epidermis of mature leaves is quite smooth. The vascular bundles are of two types, and there are flat cells which appear like oblong cavities towards the lower epidermis (pl 18 fig 1). The upper epidermis is interrupted by groups of bulliform cells (Pl 17 fig 3 Pl 18 fig 1) and the chlorophyll containing cells are arranged in loops. The bulliform cells, in a flattened leaf, is indicated by

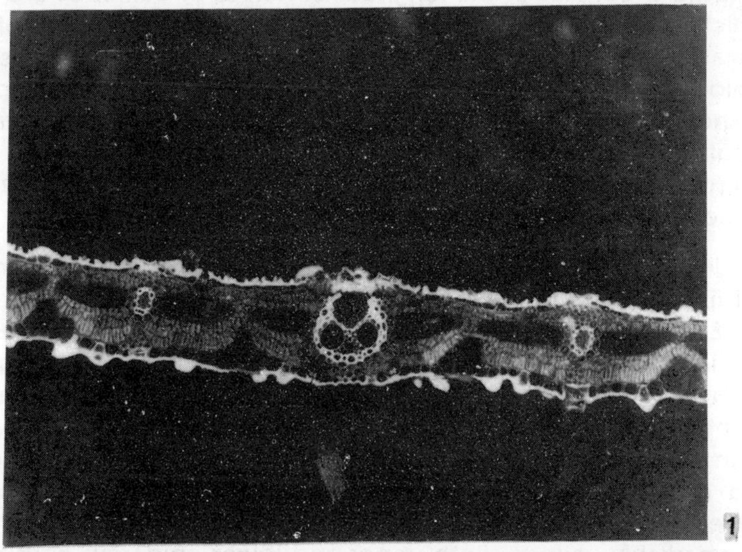

1

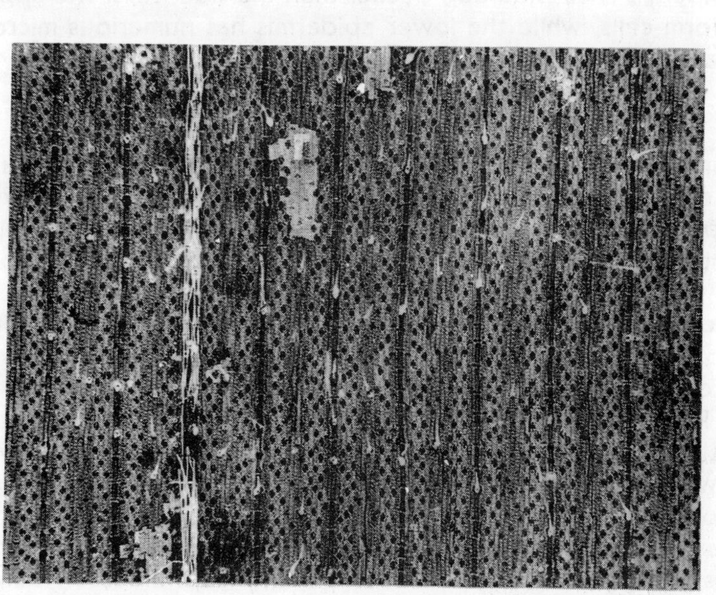

2

Pl 18. Cross section of leaf and epidermal peel of lower surface of the leaf of *Bambusa tulda*. 1. Cross section of leaf (Magnification 45X). 2. Lower epidermis of leaf (Magnification 60x)

furrows. In a large number of species they become filled with solid silica when mature. The vascular bundles in the longitudinal nerves of bamboos are enclosed in a sheath which consists of two portions. The outer portion consists of one layer of large elongated parencymatous cells with square ends which contain starch and frequently chlorophyll. In a cross section the sheath appears as a ring of orbicular cells. The inner sheath which is always present in bamboos consists of elongated cells, the end walls of which are either square or pointed. The walls are thick, strongly lignified and pitted. In bamboos the number of slender or intermediate nerves between two stout nerves varies from 1 to 10, the usual number being 5 to 8.

The midrib of the leaf always contain several vascular bundles arranged in two horizontal lines, one below the upper epidermis, and the other above the lower epidermis. Each bundle is enclosed in the girders of sclerenchymatous fibres which are placed immediately under the epidermis. The two lines of vascular bundles are separated by a band of large parenchyma cells which frequently contain starch. This tissue has intercellular spaces and the walls are often lignified. *Dendrocalamus strictus* has only one vascular bundle above and one below. But *D. hamiltonii* has 6 in the upper and 18 to 20 in the lower line. In some species the vascular bundles of the midrib are enclosed in bands of sclerenchymatous fibres. The epidermis of bamboo leaves is remarkably uniform. Both upper and lower epidermis consist of long and short cells alternating with each other. Long cells are rectangular with sinuous to wavy lateral walls and short cells are saddle shaped. The outer walls of both the upper and lower epidermis is considerably thicker than the side walls. The upper epidermis has the bulliform cells, while the lower epidermis has numerious micro and macro hairs as well as longitudinal bands of stomata (Pl. 17 fig 3 & 4; Pl 18 Fig 1 & 2)

FLOWERS

Flowers are mostly bisexual and arranged in distichous spikelets in large panicles. The spikelets consists of a number of distichous bracts or glumes. The lower and sometimes the upper glumes are empty. Each flowering glume bears in its axil the palea, a membranous transparent bract which has its back appressed to the axis of the spikelet and which is mostly two keeled and strongly ciliate except when it is terminal. Enclosed by the palea and the flowering glume is a one celled ovary with a terminal style with 2 or 3 linear plumose stigmas, surrounded by stamens which may be 3 or 6 in number occassionally more. In some genera there are also 2 or 3 membranous ciliate scales known as lodicules. The anthers of the stamens are two celled and the cells are parallel and contiguous and open longitudinally (Pl 19). The ovary structure in bamboos was studied in considerable detail by Holttum (1956) and described in his publication titled 'Classification of bamboos'. According to him in the genera *Bambusa, Thyrsostachys, Gigantochloa, Oxytenanthera, Guadua, Nastus, Dendrocalamus, Melocalamus, Pseudostachyum, Teinostachyum, Cephalostachyum, Dinochloa, Schizostachyum, Melocanna* and *Ochlandra* the ovary consists of a thin walled hollow basal part, containing the ovule, with a more or less massive top which bears the style, or the stigmas directly. The structure of the part which contains the ovule is much the same in all genera. The differences between genera are to be found mainly in the character of the top of the ovary. From the studies carried out by Arber (1954),

Holttum concluded that the origin of the bamboo ovary is from a tricarpellary condition, and presumably one with a parietal placentation from which it is derived by loss of all the ovules except one.

The top of the ovary in *Bambusa* is hairy externally and solid internally i.e. it consists of a continuous cellular tissue without any cavity in it. The only cavity is that in which the ovule hangs. In a few species of *Bambusa* the three long hairy stigmas grow directly out of the top of the ovary, but more usually the ovary top is attenuated upwards for a short distance to form a style, and the stigmas diverge from the style. In some species eg. *B. vulgaris* the style is rather long. The hairs on the ovary and style are stiff and unicellular and the hairs on the stigma are soft and multicellular. The ovary in *Dinochloa* is very small but similar to *Bambusa.*

In *Gigantochloa* and *Dendrocalamus* the top of the ovary is much the same as in *Bambusa,* but the style is always long, and usually there is only one stigma at the top of it. The style is terete, and quite slender only a short distance above the base. In *Oxytenanthera abyssinica* the hairy top of the ovary attenuates upwards very gradually into a three angled hollow structure which bears the stigmas at its apex. This structure is quite different from the style of *Gigantochloa* and *Dendrocalamus* and is more properly a part of the ovary itself because the cavity of the presumed style does not appear to be continuous with the cavity which contains the ovlue. This peculiar ovary structure is regarded by Holttum to be the distinguishing character of the genus *Oxytenanthera.*

In the genera *Pseudostachyum, Teinostachyum, Cephalostachyum, Schizostachyum, Ochlandra* and *Melocanna* the ovary is very uniform in external appearance. The top of the ovary just above the ovule-containing cavity, is somewhat widened, and then it gradually narrows upwards to form an almost terete and perfectly smooth rigid structure bearing three short stigmas at its apex. This prolongation is the style and it is a little longer than the palea at the time of flowering and so its top is clearly visible. This upward extension of the ovary is quite unlike that of *Bambusa, Gigantochloa,* and *Dendrocalamus.*

In *Arundinaria* the top of the ovary is not thickened nor swollen and hairy. Usually it bears two or three stigmas directly without a style. Also the genus *Arundinaria* has many flowered spikelets.

The other genera which have many flowered spikelets are *Bambusa, Guadua,* and *Schizostachyum.* The genus *Schizostachyum* also includes species with one flowered spikelets or more generally one perfect flower and one rudiment. A special kind of reduced condition is seen in *Gigantochloa* and *Dendrocalamus.* Here the rachilla is very short and its internodes are commonly less than 0.5mm long and it is not articulated. The lemmas and paleas of the lower florets are shorter than those of the upper ones, so that their tips occur at different levels. Such spikelets according to Holttum have as many as six flowers in a few cases, but more commonly a smaller number, and in a few cases only one. In *Arundinaria* and *Bambusa* the apical floret of the spikelet is usually imperfect and often variable in form on the same plant. The palea of the uppermost perfect floret has its back opposed to the rachilla and is two keeled. In *Dendrocalamus* the uppermost floret is usually a perfect one, but in

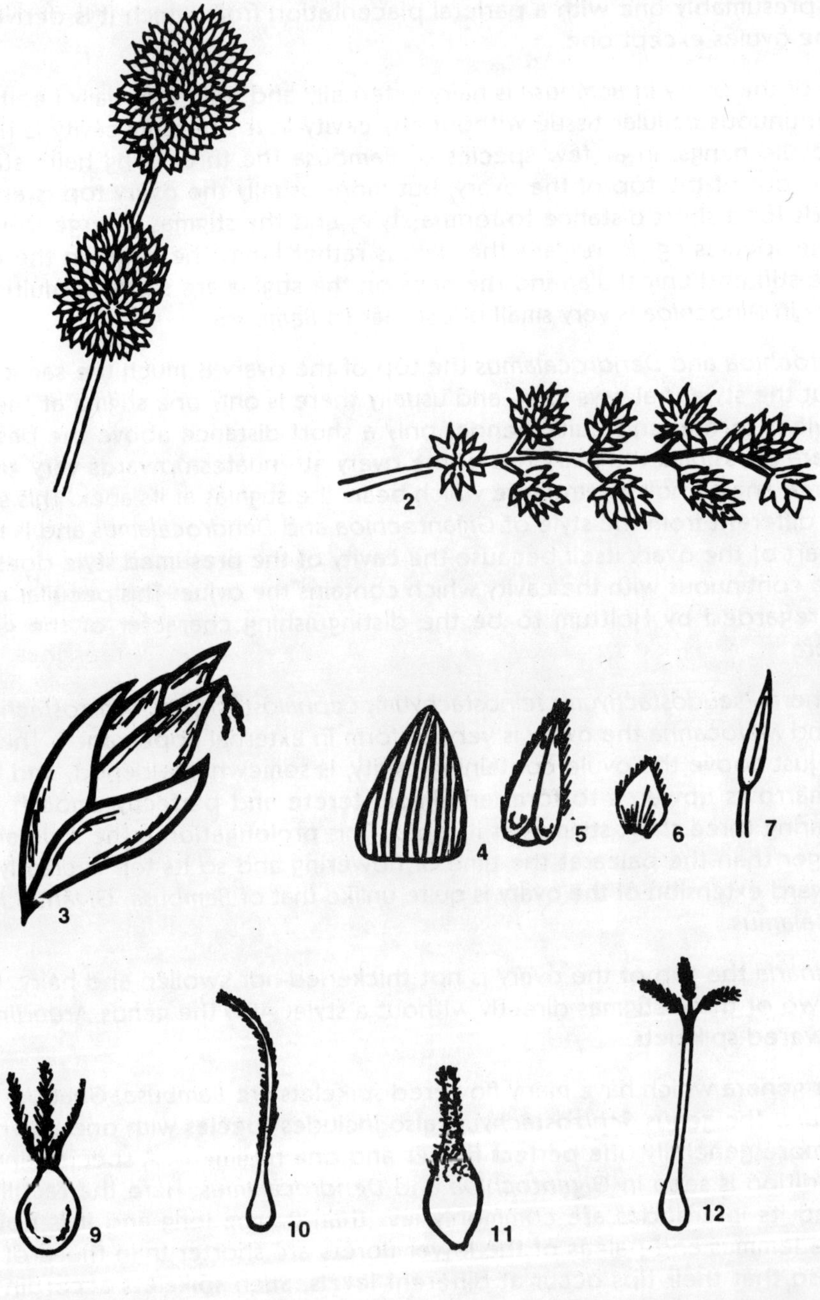

Pl1 19. Floral structure in bamboos. 1. Inflorescence of Dendrocalamus strictus. 2. Inflorescence of Bambusa arundinacea. 3. Flower 4. Lemma. 5. Palea. 6. Lodicule, 7. Stamen. 8. Pistil. 9 to 12. Different ovary structures in different bamboo species 9. Bambusa. 10&11. Gigantochloa. 12. Oxytenanthera.

some species there is a rudimentary floret above the perfect one. In the former case the palea of the uppermost perfect floret is unkeeled. If there is a rudimentary floret beyond the perfect one, the palea of the latter is more or less two keeled atleast near its apex. In the genus *Nastus* there is one perfect floret and beyond the floret an extension of the rachilla bearing a rudimentary floret. The palea of the perfect floret is two keeled. In *Chloothamnus* the spikelet and ovary structure is identical with that of *Nastus* but there is no extension of the rachilla beyond the perfect floret, and the palea of that floret lacks keels. A tube in place of separate filaments of the stamens is present in the genera *Gigantochloa, Dendrocalamus, Oxytenanthera* and *Schizostachyum*. In the genus *Dendrocalamus* only some species have a filament tube and others don't. In *Bambusa, Dendrocalamus, Gigantochloa, Thyrsostachys, Oxytenanthera, Melocalamus, Pseudostachyum, Teinostachyum, Cephalostachyum, Melocanna, Dinochloa, Ochlandra* and *Schizostachyum* the spikelets are borne in panicles which differ from the panicle of *Phragmites* mainly in having a bract and a prophyll at every branching. McClure (1934, 1935) attempted to describe the branching of the inflorescence. The branching of a spikelet tuft is difficult to understand because the first internodes of any branch are very short and buds arise very close together at the nodes of the basal part of every such branch. There is one primary branch, arising from the bud in the axil of a sheath on the main stem bearing the spikelet tufts. Each primary branch ends in a spikelet. At its base is first a two keeled prophyll, then one or more bracts with axillary buds, then usually one or more glumes which have no axillary buds, and then the lemmas with flowers in their axils. The buds in the axils of the basal bracts on the primary branch soon grow into secondary branches, and similar buds are formed at the bases of the secondary branches, which in time develop into teritary branches. As all basal buds are very close together, the result is a very close tuft of branches each ending in a spikelet.

FRUITS AND SEEDS

There are two main types of fruits in bamboos viz. the fleshy, and the caryopsis. In *Melocanna* for instance the pericarp is thick and fleshy and the fruit is more or less pear shaped tapering to the remains of the style. The seed is almost entirely composed of the scutellum which during development of the seed absorb all the endosperm except a thin film separating the embryo from the pericarp. The seed of *Melocanna* germinates while the fruit is still attached to the parent plant.

The fruits of *Melocalamus compactiflorus* are very different in appearance from those of *Melocanna*. When dried whole for preservation in the herbarium, they are more or less spherical, with diameter 1.5 to 2.5 cm. After soaking in water the pericarp of such a specimen becomes soft and fleshy but it does not separate readily from the seed, nor is it nearly so thick in proportion to the size of the seed as in *Melocanna*. Fruits with pericarp thick and fleshy to the base are found in the genera *Melocalamus, Melocanna, Ochlandra,* and *Dinochloa*. But these fruits are developed from ovaries of different kinds (Holttum 1956).

In most of the rest of the genera the fruit has a thin membranous or crustaceous pericarp and is a caryopsis. The fruits of *Bambusa, Gigantochloa,* and *Dendrocalamus* are very similar. Further the ovary and fruit structure of *Pseudostachyum, Teinostachyum,*

Cephalostachyum and *Schizostachyum* are very similar. Some genera eg. *Dinochloa* and *Ochlandra* have no endosperm in the ripe seed.

GENERIC CHARACTERS OF SOME BAMBOOS

1. *ARUNDINARIA*

Culms are tufted, erect, rarely single stemmed and occur in temperate regions. Culms are mostly less than 6 metres high. Leaves are small with transverse veins conspicuous in most species, dividing the leaf into rectangles or squares. Nerves in quarter inch 21 to 36 numbers. Inflorescence in racemes or panicles, and spikelets often pedicelled and one to many flowers on each spikelet. Empty glumes 1 to 2, palea 2 keeled, lodicules 3 and ciliated. Stamens usually 3, style short, stigmas 2 or 3 plumose. Pericarp of fruit thin, membranous and adnate to the seed.

Species of *Arundinaria* are differentiated on the basis of the presence or absence of thorns at the nodes, bracts of inflorescence large or small, presence or absence of leaves during flowering, and transverse veins of leaves prominent or not so *prominent.*

2. *BAMBUSA*

Culms tall, erect and tufted. In a few species eg. *Bambusa nutans* and *B. khasiana* they are nearly single stemmed. Transverse veins on leaves *not* conspicuous. Nerves generally 20 to 40 on a quarter inch. Exceptions are *Bambusa nana* (45-54), *B. polymorpha* (40-45), *B.binghami* (42-45), *B. khasiana* and *B. griffithiana* (18-21), spikelets sessile, palea entire or slightly 2-dentate, all prominently two keeled and not cleft. If cleft only slightly. Stames 6, filaments free. Ovary oblong or obovate, tip of ovary hairy. Fruit a caryopsis grooved on one side, pericarp thin, and adherent to the seed.

Species of *Bambusa* are differentiated on the basis of the nature of spikelets, number of fertile flowers, presence or absence of auricles in the culm sheath, and branching of the culm.

3. *CEPHALOSTACHYUM*

Culms tall and tufted. In most species the stouter nerves of leaves very broad. Transverse veins often visible but not conspicuous. Infloresence in dense heads broadly obconical or nearly globose consisting of numerous short spikes, each spike with several spikelets. The lowest spikelets usually sterile with numerous empty glumes, the uppermost spikelet with one fertile and several empty glumes. Paleae thin convolute, many nerved, the two keels close together, stamens 6. Fruit a small caryopsis glabrous, beaked and supported by the persistent glumes and lodicules, and embedded in a densely packed mass of dry sterile spikelets. The various species of *Cephalostachyum* are differentiated by details of inflorescence, culmsheaths, leaves, and the branching of the culms.

4. *DENDROCALAMUS*

Culms tall and unarmed in dense clumps, culm sheaths often very large. Leaves mostly large, nerves 20 to 30 on quarter inch, in most species 24 or 25. Exceptions are *Dendrocalamus membranaceous* 30 to 40. Transverse veins if not obscured by hairs more or less visible in dry leaves on the underside. Spikelets ovate, flowers few, rarely more than 6, usually bisexual. Glumes many nerved, empty glumes 2 to 3. Palea of lower flowers keeled and ciliate. Only those of the terminal flowers are rounded on the back and not keeled. Lodicules absent but occasionally 1 to 2 in *D. patellaris, D. brandisii* and *D. flagellifer,* Stamens 6, filaments free. Ovary hairy in the upper part, style long hairy and undivided. In a few species 2 to 3 fid at the apex. Seed enclosed in a hard crustaceous pericarp. Different species are differentiated by the number, nature, and arrangement of the spikelets.

5. *DINOCHLOA*

Evergreen, climbing, culms zigzag, geniculate. Transverse veins often visible on the underside of dry leaves. Spikelets minute one-flowered, with one to four empty glumes, palea convolute, not keeled, lodicules absent. Fruit ovoid, and mucronate. Two types of fruits occur in the different species (i) a small fruit with a firm smooth pericarp and a normal gramineous seed consisting mostly of endosperm (ii) a large spherical fruit with a fleshy pericarp and a seed which appears to consist almost entirely of scutellum.

6. *GIGANTOCHLOA*

Culms tall, erect, scandent, and branchless in the lower portion. Leaves pale and hairy beneath. Inflorescence consist of long fertile and short sterile spikelets in heads. Fertile spikelets linear, and sharply subulate. Fertile flowers 2 to 4, Glumes ciliate with a conspicuous black fringe on the edge, and a few black stiff hairs on the back in *G. macrostachya.* Palea of all flowers 2-keeled and keels ciliate. Stamens 6, filaments connate and united into a tube, ovary hairy. Fruit a caryopsis narrowly cylindric, tipped with a persistent style. Pericarp thin, membranous, and adnate to the seed.

7. *MELOCANNA*

Culms distant, erect, and the lower two thirds of the culm bare of branches. Culm sheaths coriaceous, persistant, brown in colour in contrast with the bright green of the internodes. Upper one third of culm sheath wavy and not appressed to the culm, sparsely hairy outside, top concavely truncate with rounded auricles, blade recurved, narrow longer than the sheath, and narrowed into a long convolute apex, glabrous and striate on both sides. Leaves glaucous, glabrous, or with a few scattered hairs near the base, nerves 18 to 24 on quarter inch, inner edge closely set with fine hairs on a broad colourless band. Transverse veins visible, with evenly distributed pellucid dots. Spikelets usually 2 to 5 on an arrested axis so as to appear fascicled, in the axils of large bracts. Each spikelet with numerous empty glumes and one flower rarely bisexual, usually male with a rudimentary pistil, or female with sterile anthers. In the axils of the lower bracts are the buds which develop into short branchlets with a spikelet at the end. The fascicles of spikelets are on unilateral spikes arranged in long narrow pedunculate panicles. The panicles are of unequal length and most of the long ones are androdynamous, while the shorter ones are gynodynamous. Bisexual flowers occur occasionally in both types of panicles. In some the spikelet is

2-flowered. The palea of the lower flower has one or two pairs of nerves. Fruit is large, fleshy, pyriform and beaked. Seed when ripe has no endosperm, but the embryo is with a large scutellum.

8. *MELOCALAMUS*

Loosely tufted climber, culms nearly solid, culmsheath thick coriaceous, shorter than the internodes with the blade linear-lanceolate and recurved. Nerves on leaves 20 to 30 on quarter inch. Flowers frequently. Panicles large and often leafy. Branches fistulose bearing half whorls of long hairy spikes with globose compact flower heads about 1cm in diameter. Spikelets small and as long as broad, fertile flowers 2 in number, lodicules large and ciliate, ovary glabrous, style short, stigmas 2 or 3. Fruit 2.5 to 3.0 cm in diameter, and fleshy while immature, and later coriaceous. Embryo has a large scutellum without endosperm.

9. *NASTUS*

This genus occurs mainly in the southern hemisphere from Madagascar to Solomon Islands. Culms are thin walled scrambling and scandent with many branches at each node. Each spikelet has one perfect floret with several empty glumes of increasing size below it. Beyond the perfect floret an extension of the rachilla bears a rudimentary floret. The palea of the perfect floret is two keeled. The vegetative parts of this genus resemble very closely those of *Racemobambos* and the spikelet structure is identical with that of *Chloothamnus* the only difference being that there is no extension of the rachilla beyond the perfect floret and the palea of the floret lacks keels. Otherwise even in ovary structure the genera *Nastus* and *chloothamnus* resemble each other.

10. *OCHLANDRA*

Culms tufted, gregarious. Culm sheath thin and persistant. Spikelets mostly arrested in their growth remains small, with a few large fertile ones consisting of 2 to 6 empty and one flowering glume. The glumes bear conspicuous transverse veins. Palea thinly membranous, convolute often emarginate, longitudinal nerves numerous, two of them generally stronger than the rest. Lodicules 2 to 12, stamens 6 to 120, filaments free, more or less connate, anthers long and exserted, linear and usually mucronate. Style elongated, stigmas 4 to 8 linear, plumose, usually closely twisted before expanding. Fruit beaked, large, supported by the persistent glumes, pericarp thick, fleshy and filled with starch. Endosperm in the immature fruit soft and fleshy. Some species flower annually, and others at long intervals and the inflorescence always bear leaves.

11. *OXYTENANTHERA*

Culms tall and erect. Spikelets few flowered. Palea of upper flowers indistinctly or not at all keeled, palea of the lower flowers 2-keeled. Stamens 6, filaments united into a thin membranous tube. The ovary is hairy at the top and attenuates upwards very gradually into a more or less three angled hollow structure which bears the stigmas at its apex. The cavity of the style is not continuous with the cavity that contains the ovule. Holttum regards this peculiar structure of the ovary to be a distinguishing character of the genus *Oxytenanthera*. Inflorescence is often leafy. Many species flower frequently and some annually. Fruit is a caryopsis grooved and terminated by a beak, pericarp thin membranous and adnate to the seed. Species of *Oxytenanthera* are distinguished on the basis of the size of panicles, and arrangement of spikelets.

12. *PHYLLOSTACHYS*

This genus occurs mostly in the temperate regions. Culms are less than 6 metres high and tufted. Rhizomes have long creeping branches which sends up tufts of culms at a distance from the parent clump. Branches on the culm are flattened on the inside above the axillary bud and hence angular. Branches 2 to 3 on each node. Leaves tessellated. Transverse veins of the leaves always conspicuous and close together usually dividing the leaves into minute squares. Panicles terminal and leafy, spikelets sessile supported by prominent sheathing bracts, often with a leafy blade. Flowers 1 to 4 on each spikelet in the axils of large imbricate bracts often with a small leafy blade. Empty glumes 2 to 3, palea 2 keeled many nerved often bimucronate, stamens 3, style long with three long feathery stigmas. Species differentiated on the basis of size of culms, presence or absence of auricles on the culm sheaths and texture of leaves.

13. *PSEUDOSTACHYUM*

Culms single stemmed often overhanging and supported by neighbouring trees. Lower part branchless. Culms thin walled and culm sheath loose, much shorter than the internodes, blade early deciduous, ribbed and often with transverse veins, lanceolate or triangular often longer than the sheath. Leaves thin, large, usually glabrous, sometimes fine hairs on the underside. Transverse veins conspicous often bent where they cross the band of silica cells. Flowers frequently. In most cases flowers are deformed and sterile, appearing as large round masses of hairy glumes. Spikelets solitary in the axils of narrow bracts on the slender branches of large leafy panicles. One fertile flower, palea convolute 2-keeled, keels not ciliate, lodicules large and ciliate, anthers apiculate. Fruit a caryopsis, depressed globose, surmounted by the base of style, and supported by the persistent glumes, palea, and lodicules.

14. *SCHIZOSTACHYUM*

Culms tufted, overhanging and supported by trees, culm walls thin, culm sheaths much shorter than internodes, and thin with a broad base and tapering top and two small auricles at the apex, blade narrow and reflexed and as long as the sheath. Leaves large with fine hairs on the underside, nerves on the leaves about 21 on a quarter inch, transverse veins prominent oblique and bent. Inflorescence a long spike terminating in leafy branchlets with distant half whorls. Spikelets one-flowered in some species, many flowered in others. More usually it has one perfect flower and one rudiment. The fertile spikelet longer than the sterile spikelet. Empty glumes 2 to 4, palea convolute, minutely 2-dentate, keels indistinct with elongated points at the apex. Lodicules 3 and unequal. Anthers yellow and obtuse, ovary glabrous continued into a thick cylindric hollow style terminated by 3 long plumose stigmas. Fruit is a caryopsis ellipsoid-cylindric crowned by the long persistent style.

15. *TEINOSTACHYUM*

Culms thin, overhanging and sometimes climbing. Spikelets upto 8.0 cm long, slender many flowered, sometimes pedunculate in bracteate whorls on branches of a leafy panicle, upper and lower florets imperfect, rachilla elongate between the flowers, glumes mucronate, palea convolute, keeled, keels ciliate near the apex. Lodicules 3, nerves 3-9. Style enclosed in, and partly adnate to, a cylindric sheath. Fruit a caryopsis ovoid, beaked, with a crustaceous pericarp.

16. *THYRSOSTACHYS*

Culm tall, erect, and tufted. Inflorescence in panicles consisting of numerous pedunculate spikes in the axils of sheathing bracts, peduncles bearing coriaceous sheaths, usually without a blade. Branches of compound spikes short, in the axils of sheathing bracts, bearing perfect and imperfect spikelets in the axils of membranous sheaths. Palea hyaline, keeled, ciliate along the keels, those of the lower flower in each spikelet deeply bifid, teeth awned, and that of the uppermost flower entire or nearly so and indistinctly keeled. Stamens 6, filaments free. Fruit a caryopsis, long cylindric, beaked, pericarp thin, membranous, and adnate to the seed.

The two species in the genus are differentiated on the basis of (i) the nature of the culm sheaths which is persistent, tightly appressed to the culm, half to three fourths the length of the internode, and blade caducous in *T. oliveri* while in *T. siamensis* culm sheaths are deciduous, as along as the internodes, and blade is narrowly triangular. (ii) Leaves in *T. oliveri* are large being 17 to 23 cm in length and 1 to 3 cm in width, with 27-36 nerves on quarter inch. Sheaths keeled and minutely ciliate along the edges, while in *T. siamensis* leaves are smaller being 7 to 15 cm in length and 0.8 to 1.25 cm in width and nerves 33 to 48 on quarter inch.

In addition, three new genera of Chinese bamboos were described by Wen Taihui and these are *Gelidocalamus, Clavinodum*, and *Ampelocalamus.*

17. *GELIDOCALAMUS* Wen

Shrubby with amphipodial rhizomes, internodes of culms cylindrical, unfurrowed, with 7 to 12 branches at each node. Branches short and slender, simple having 2 to 3 nodes, usually with one leaf on it, and the branch sheath is longer than the internode. Culm sheaths are persistent, auricles rudimentary or lacking, ligule extremely short, curved or wedge shaped. Leaf blades broadly lanceolate, acute at the apex, and decurrent. Transverse veinletes obvious on both sides. Inflorescence paniculate. spikelet very small, with 3 to 5 florets. The floret has 2 glumes, lemma with keels accuminate and shorter than palea. Lodicules 3, stamens 3, style 2.

18. *CLAVINODUM* Wen

Shrubby or subarborescent, amphipodial rhizomes, internodes of culms cylindrical, nodes rigid, irregular, culm sheaths persistent or late dropping, much shorter than the internode, auricles well developed sometimes absent, sheath blades subulate and recurved, branches 3 to 5 at a node and slender. Inflorescence determinate, and racemose, lateral, 3 spikelets in each. Spikelets with 2-7 florets but generally only one floret can be seen on a spikelet because the rachilla is caducous. Spike petiole slender, zig zag, palea tips bilobed; lodicules 2, sometimes 3, stamens 3 to 4, filaments free, styles 2 sometimes 3.

19. AMPELOCALAMUS S.L. Chen Wen et C.Y. Sheng

This is a peculiar genus and differs from other genera of bambusoideae by a number of characters. Rhizome sympodial, culms slender, climbing, branches fine and numerous, culm sheaths much shorter than the internodes. Both culm sheath auricles and leaf sheath auricles well developed, numerous, long, setose, radiating at the margin, sheath ligules long, and ciliate at the apex, leaf blades setose. Inflorescence determinate racemes. Spikelet petioled with 5 to 7 florets, stamens 3, styles 2.

20. *SHIBATEA*

Culms short and slender grooved on one side at the base of the internodes. Culm sheaths caducous. Leaf sheath undeveloped. 3 to 5 short branches at each node.

21. *CHIMONOBAMBUSA*

Densly leafy culms. Nodes not prominent. Buds numerous at the nodes of culms. Culm sheaths shorter than internodes. Leafy sheaths deciduous, bristles *not* scabrid. Spikelets racemose, palea 2 keeled, stamens 3, pericarp thin and adnate to the seed.

22. *SEMIARUNDINARIA*

Rhizomes monopodial and nonclump forming, culms erect and medium sized. Culm sheaths caducous. 3 to 5 branches at each node. Branches cylindrical without grooves. Culm sheaths ciliate and caducous. Inflorescence a simple spikelet.

23. *SINOBAMBUSA*

Rhizomes monopodial and nonclump forming. Culms erect. Long purplish hair at the nodes of new culms. Internodes of culms long. Culm sheaths ciliate. Branches arise from a triple branching bud at the nodes, and the three buds elongate simultaneously into 3 branches. Side branches longer than one half of its main branch. Inflorescence a simple spikelet. Outer glumes present.

24. *TETRAGONOCALAMUS*

Square culms, aerial roots at the lower nodes. Culm sheaths caducous without appendages. 3 branches at each node and each branch is independent upto to base.

25. *INDOCALAMUS*

Culms elongated, tufted and much branched. Nodes rather enlarged. Buds solitary at the nodes of the culm. Leaf sheaths persistent, Bladeless, bristless scabrid or setulose. Culm sheaths papery. Spikelets panicled, palea 2 keeled, stamens 3, pericarp thin and adnate to the seed.

Makino who was the first to study Japanese Bamboos established a number of new genera between the years 1901 to 1944 viz. *Sasa, Shibatea, Chimonobambusa,*

Semiarundinaria, Sinobambusa, Pseudosasa, Sassella, Pleioblastus, Sasamorpha, Tetragonocalamus, and *Nipponocalamus.*

In *Pseudosasa, Sasamorpha* and *Sassella* the prophyll is of the "Sasa type" i.e. the prophyll is of a two keeled form and the culms produce only one branch at the nodes. The genera *Semiarundinaria, Sinobambusa, Pleioblastus* and *Nipponocalamus* have the Arundinaria type of prophyll i.e. they have atleast three two keeled prophylls and produce atleast three branches at each node. Though some have seven to nine branches at the nodes, they develop from secondary buds. The genus *Tetragonocalamus* has the chimonobambusa type of prophyll i.e. there are three branches at each node and they are characterized by three separate prophylls, and the three branches are independent upto the base. The genus *Semiarundinaria* is distinguished by a long bract which covers the spikelet.

5

SHORTCOMINGS OF BAMBOO CLASSIFICATION

REASONS FOR SHORTCOMINGS

In spite of the tremendous amount of research carried out so far on bamboos, their classification is far from a final solution. The main reason being the nonavailability of material required for taxonomical studies, and when available, only material from a limited range of their distribution could be obtained by the researchers, and even that peicemeal. Thus research has been conducted by various researchers working on different parts of the plant - some on rhizomes, some on leaves, some on culms, some on culm sheaths and some on flowers from herbarium sheets. As such, the scope for errors are many. For instance Brandis points out that *Bambusa falconeri* Munro was based on the leaves of *Dendrocalamus hamiltonii* and flowers of *Bambusa nutans.*

Holttum (1956) points out that the type species *Dinochloa tjankorreh* (Schult) Buse from Java has an earlier name *Bambusa scadens* Bla. In 1928 Backer uses the name *Dinochloa scadens* (B1) kuntze for the same species. Steenis looked over many herbarium specimens of this species in Leiden herbarium, and found that it agreed with Backer's description. But the description also fits the description of a fruit of *Dinochloa andamanica* Kurz at Kew. Moreover the fruit also had similarities with the fruit of *Dendrocalamus strictus* differing only in having a larger embryo. Furthermore there is a climbing bamboo in Borneo and the Philippines called *Dinochloa scandens* by Gamble which has quite a different type of fruit.

Holttum also calls attention to another case where the fruit of *Schizostachyum acutiflorum* Munro now called *Schizostachyum diffusum* (Blanco) Merril, was earlier called *Dinochloa diffusa* by Merril, is externally like that of the Philippine and Bornean bamboo called *Dinochloa scandens* by Gamble.

In another instance Brandis points out that Kurz in Forest Flora, Volume II, states that *Bambusa affinis* Munro grows in the forests of Martaban east of Sitang River. But Gamble identifies this species with a low bush growing in the Calcutta Botanic Garden on the edge of a muddy river bank. The specimens of this species in Kew Herbarium however have entirely different leaves about 3.0 cm wide with 18 to 21 nerves on a quarter inch. The leaf sheath is sharply keeled and tightly appressed to the internode and the ligule is large.

Brandis points out another case of confusion in the names of bamboo species where *Bambusa pallida* Munro which is indigenous in Sikkim and Bhutan upto 1800 metres and also in Khasi Jaintia and Naga hills is identified by Gamble in his Bambuseae as *Dendrocalamus criticus* Kurz.

There are also confusion in names of bamboos due to revision of taxonomic position eg. *Oxytenanthera nigrociliate* Munro was earlier known as *Oxytenanthera auriculata* Prain, *Gigantochloa andamanica* Kurz, also *Gigantochloa auriculata* Kurz, and *Bambusa auriculata* Kurz. Another example is *Ochlandra scriptoria* was earlier known as *Ochlandra rheedii.* Also *Ochlandra sivagiriana* was known as *Ochlandra rheedii* var *sivagiriana. Ochlandra talbotii* was also known as *Ochlandra rheedii* var sivagiriana Talbot.

Some Japanese Researchers have revised the genera *Arundinaria* and proposed eight genera in place of *Arundinaria* Munro. The characters used by Nakai and others to delimit the genera are vegetative characters and may therefore require further investigation preferably by modern methods of finger printing before their proposal is confirmed.

INITIAL STEP IN REVISION OF BAMBOO TAXONOMY

In order to revise the taxonomy of bamboos it would however be necessary to index all the names of bamboo species and then sort out the synonymns and then correct their taxonomic position. As research on bamboos are being carried out on various aspects and all researchers do not always refer to taxonomists for the correctness of the name, some of the synonymns are still being used as the name of the species, and often the botanical name is given without citing the authority. As such an attempt is made here to list in alphabetic order all the names of bamboo species currently appearing in research publications and the countries of their occurrence (Table 10).

In fact Ohrenberger and Goerrings 1989 have made a start in this direction by preparing a documentary survey of existing bamboos the world over in their publication titled "Bamboos of the World". They have made a tentative list of bamboos known in different parts of the world along with their geographical distribution and published literature on each species. Though it is not a taxonomical revision, they have attempted a classification of bamboos into family, subfamily, tribe, subtribe, genera and species. In the case of each genus, botanical names are cited indicating their status as genus, subgenus, section, species, subspecies, variety, form or cultivar. A list of synonymns is also provided. Descriptions of the plants however have not been given. They have mentioned that this publication is a bibliographic treatment without herbarium and field research. The usefulness of this publication lies in the ready availability of information on the names by which various bamboos are known and their distribution.

Table 10. BAMBOO SPECIES IN ALPHABETIC ORDER AND THE COUNTRIES OF THEIR OCCURRENCE.

Sl. No.	Name of Species	Countires of Occurrence
1.	*Ampelocalamus actinotrichus* (Merr and Chun) S.Z. Chen, Wen et G.Y. Sheng	China

2.	*Arundinaria alpina* K. Schun	E. Africa (Kenya)
3.	*Arundinaria ambilis* McClure	China
4.	*Arundinaria anceps*	India
5.	*Arundinaria aristata* Gamble	India
6.	*Arundinaria armata* Gamble	India
7.	*Arundinaria callosa* Munro	India
8.	*Arundinaria ciliata* A. camus	Thailand
9.	*Arundinaria clarkei* Gamble	India
10.	*Arundinaria deblis*	Srilanka
11.	*Arundinaria densifolia* Munro	India, Srilanka
12.	*Arundinaria elegans* Kurz	India
13.	*Arundinaria falconeri* Benth & Hook F.	India
14.	*Arundinaria falcata* Nees.	India
15.	*Arundinaria floribunda*	Sri Lanka
16.	*Arundinaria fortunei*	E. Africa
17.	*Arundinaria gallatlyi* Gamble	Burma
18.	*Arundinaria gigantea*	Bangladesh
19.	*Arundinaria gracilis* Munro	India
20.	*Arundinaria griffithiana* Munro	India
21.	*Arundinaria hirsuta* Munro	India
22.	*Arundinaria hookeriana* Munro	India
23.	*Arundinaria intermedia* Munro	India
24.	*Arundinaria invatekensis*	China
25.	*Arundinaria japonica* Sieb & Zucc ex Steud	Indonesia
26.	*Arundinaria jaunsarensis* Gamble	India
27.	*Arundinaria khasiana* Munro	India
28.	*Arundinaria kurzii* Gamble	India
29.	*Arundinaria maling*	Nepal, India
30.	*Arundinaria mannii* Gamble	India
31.	*Arundinaria microphylla* Munro	India
32.	*Arundinaria mitskayamensis*	Japan
33.	*Arundinaria muculata*	China
34.	*Arundinaria patlingii* Gamble	India
35.	*Arundinaria polystachya* Kurz	India
36.	*Arundinaria prainii* Gamble	India
37.	*Arundinaria pusilla* A. Chev & A. Camus.	Thailand
38.	*Arundinaria racemosa* Gamble	India, Nepal
39.	*Arundinaria rolloana* Gamble	India
40.	*Arundinaria scandens*	Sri Lanka
41.	*Arundinaria simonii* (Carr) Reviere	Japan
42.	*Arundinaria suberecta* Munro	India, Thailand
43.	*Arundinaria spathiflora* Trinius	Indonesia
44.	*Arundinaria suberosa*	India
45.	*Arundinaria tecta*	Bangladesh
46.	*Arundinaria walkeriana* Munro	India, SriLanka

47.	*Arundinaria wigthiana* Nees	India
48.	*Bambusa affinis* Munro	Malaysia
49.	*Bambusa arundinacea*, willd,	India, Thailand, Malaya, Indonesia, Philippines.
50.	*Bambusa atra* Lindl	India, Indonesia
51.	*Bambusa amahussana*	Malaysia, Philippines
52.	*Bambusa auriculata*	India
53.	*Bambusa balcooa Roxb*	India, Indonesia
54.	*Bambusa bambos* Becker	Sri Lanka, Indonesia
55.	*Bambusa bambusoides*	Philippines
56.	*Bambusa bambusoides* Var. aurea Makino	Philippines
57.	*Bambusa biciatricatus*	China
58.	*Bambusa basihirsuta* McClure	China
59.	*Bambusa binghamii* Gamble	Malaya
60.	*Bambusa blumeana* Bl. ex Schultf.	Thailand, Malaya, Indonesia, Philippines.
61.	*Bambusa brassii*	Malaya
62.	*Bambusa brevicephala*	Malaya
63.	*Bambusa breviflora* Munro	China
64.	*Bambusa burmanica* Gamble	India, Burma, Malaya.
65.	*Bambusa chungii*	China
66.	*Bambusa copelandi* Gamble	India, Burma.
67.	*Bambusa cornuta* Munro	Philippines
68.	*Bambusa dissemulater*	China
69.	*Bambusa dissemulater* var. albonida	China
70.	*Bambusa dolichoclada*	Singapore
71.	*Bambusa dolichomerithala* Hayata	China
72.	*Bambusa emeiensis*	China
73.	*Bambusa eutudoides* McClure	China
74.	*Bambusa flexuosa*	Malaya
75.	*Bambusa floribunda* Nakai	Philippines
76.	*Bambusa forbesii*	Indonesia
77.	*Bambusa fruiticosa*	Indonesia
78.	*Bambusa gaudua*	Costa Rica
79.	*Bambusa gibbodes* Lin.	China
80.	*Bambusa glaucescens* (willd) *Sieb ex Munro.*	China, India, Malaya, Indonesia, Singapore.
81.	*Bambusa glaucescens* Var. shimata (Hayata) Chia ex But.	China
82.	*Bambusa griffithiana* Munro	India
83.	*Bambusa heterostachya*	Malaya, Singapore.
84.	*Bambusa hirsuta*	India
85.	*Bambusa hiretellus*	Malaya
86.	*Bambusa horsfeldii* Munro	Indonesia

87.	*Bambusa khasiana* Munro	India
88.	*Bambusa kingiana* Gamble	India
89.	*Bambusa klosii*	Malaya
90.	*Bambusa lapidea* McClure	China
91.	*Bambusa lineata* Munro	India
92.	*Bambusa longispiculata* Gamble	India
93.	*Bambusa magica*	Malaya
94.	*Bambusa marginata* Munro	Malaya
95.	*Bambusa mastersii* Munro	India, Malaya
96.	*Bambusa merrillii* Gamble	Philippines
97.	*Bambusa microcephala*	Malaya
98.	*Bambusa montana*	Malaya
99.	*Bambusa multiplex* (Lour) Raeusch.	China, Japan, India, Srilanka, Indonesia, Philippines.
100.	*Bambusa multiplex* Var. *Nana* (Roxb) Keng f.	China
101.	*Bambusa nana* Roxb.	Thailand, China, Philippines.
102.	*Bambusa nutans Wall ex* Munro	India
103.	*Bambusa oldhami* Munro	China
104.	*Bambusa oliveriana* Gamble	Burma, India
105.	*Bambusa orientalis*	India
106.	*Bambusa pachinensis*	China
107.	*Bambusa pachinensis* Var. *hirsutissima* (odash) Lin.	China
108.	*Bambusa pallida* Munro	India
109.	*Bambusa pauciflora*	Malaya
110.	*Bambusa pergracile*	Singapore
111.	*Bambusa pervariabilis* McClure	China
112.	*Bambusa pierrie* E.G. Camus.	Thailand
113.	*Bambusa pierreana*	Philippines
114.	*Bambusa polymorpha*	India, Thailand, Indonesia.
115.	*Bambusa prasina* Wen	China
116.	*Bambusa ridleyi*	Malaya, Singapore
117.	*Bambusa rigida*	China
118.	*Bambusa riparia*	China
119.	*Bambusa rumphiana* Kurz	China
120.	*Bambusa tutila*	China
121.	*Bambusa schizostachys*	India
122.	*Bambusa schizostachyoides* Kurz	India
123.	*Bambusa solomensis*	Solomon Island
124.	*Bambusa spinosa Bl.*	India, Indonesia
125.	*Bambusa sinospinosa*	China
126.	*Bambusa subtrincata* chia ex Fung	China

127.	*Bambusa textilis* McClure	China
128.	*Bambusa textilis* Var. *albostricta* McClure	China
129.	*Bambusa textilis* Var. *glabra* McClure	China
130.	*Bambusa teres* Ham	India, Singapore
131.	*Bambusa tulda* Roxb	India, Thailand, Burma, Indonesia, Singapore, Philippines.
132.	*Bambusa tuldoides* Munro	China
133.	*Bambusa variegata*	Singapore
134.	*Bambusa variostriatus*	China
135.	*Bambusa ventricosa* McClure	China, India, Malaya, Philippines, Singapore.
136.	*Bambusa verticillata*	Singapore
137.	*Bambusa villulosa*	
138.	*Bambusa vulgaris* Schrad	India, Srilanka, Malaya, Indonesia, Philippines, China, Bangladesh, New Guinea.
139.	*Bambusa vulgaris* Var. *striata* (Lodd) Gamble	Philippines
140.	*Bambusa vulgaris* Var. *vittata*	Philippines
141.	*Bambusa wamin* Brandis	Thailand.
142.	*Bambusa wrayi*	Malaya
143.	*Bashania fargesi* Keng fet yi	China
144.	*Cephalostachyum burmanicum*	Burma
145.	*Cephalostachyum capitatum* Munro	India
146.	*Cephalostachyum capitatum* Var. *decomposita*	Sikkim, India
147.	*Cephalostachyum flavescens* Kurz	India
148.	*Cephalostachyum fuschianum* Gamble	India
149.	*Cephalostachyum latifolium* Munro	India, Sikkim
150.	*Cephalostachyum mindorense* Gamble	Philippines
151.	*Cephalostachyum pallidum* Munro	India
152.	*Cephalosatchyum pergracile* Munro	India, China, Thailand.
153.	*Cephalostachyum virgatum* Kurz	Thailand
154.	*Chimonobambusa armata* (Gamble) Hsueh ex yi	China
155.	*Chimonobambusa callosa*	India
156.	*Chimonobambusa convoluta* Dai ex Tao.	China
157.	*Chimonobambusa callosa*	India

158.	*Chimonobambusa densifolia*	India
159.	*Chimonobambusa falcata*	India
160.	*Chimonobambusa griffithiana*	India
161.	*Chimonobambusa hookeriana*	India, Sikkim.
162.	*Chimonobambusa intermedia*	India, Sikkim.
163.	*Chimonobambusa jaunsarensis*	India
164.	*Chimonobambusa khasiana*	India
165.	*Chimonobambusa marmorea* (Mitford) Makino	Japan
166.	*Chimonobambusa polystachya*	India
167.	*Chimonobambusa quadrangularis* (Feuzi) Makino.	China
168.	*Chimonobambusa setiformis* Wen	China
169.	*Chimonocalamus pattens* Hseuh ex yi	China
170.	*Chimonocalamus fimbricatus* Hseuh ex yi	China
171.	*Chloothamnus* products Pilger	China
172.	*Clavinodum globinodum* (C.H.Hu) Keng fet Wen.	China
173	*Clavinodum oedogonatum* (Z.P. Wang et G.H. Ye) Wen.	China
174.	*Chusquea species*	Central and South America
175.	*Davidsea attenuata*	Sri Lanka
176.	*Dendrocalamus asper* Back	India, Sri Lanka, Thailand, Bangladesh, Malaya, Indonesia.
177.	*Dendrocalamus brandisii* Kurz.	India, Burma, China, Thailand.
178.	*Dendrocalamus calostachyus* Kurz	Burma, India.
179.	*Dendrocalamus cinctus*	Srilanka
180.	*Dendrocalamus cirranii* Gamble	Philippines
181.	*Dendrocalamus colleltianus* Gamble	Burma, India
182.	*Dendrocalamus criticus* Kurz	India
183.	*Dendrocalamus dumosus* (Ridl) Holttum.	Malaya, Thailand.
184.	*Dendrocalamus elegans*	Malaya, Thailand.
185.	*Dendrocalamus flagellifer* Munro	Philippines
186.	*Dendrocalamus giganteus* Munro	India, Burma, Sri Lanka, China, Thailand, Malaya, Indonesia.
187.	*Dendrocalamus hamiltonii* Nees & Arn.	India, Thailand.
188.	*Dendrocalamus hirtellus*	Thailand, Malaya.
189.	*Dendrocalamus hookeri* Munro.	India, Nepal
190.	*Dendrocalamus hookeri* Var. *parishi*	India
191.	*Dendrocalamus latiflorus* Munro	China, Philippines
192.	*Dendrocalamus longifimbricatus* Gamble.	India

193.	*Dendrocalamus longispathus* (Kurz) Benth.	India, Thailand
194.	*Dendrocalamus membranaceus* Munro.	India, Burma, Thailand, Sri Lanka,
195.	*Dendrocalamus merrillianus* (Elm) Elm.	Philippines.
196.	*Dendrocalamus minor*	China
197.	*Dendrocalamus pattellaris* Gamble.	China, India, Nepal
198.	*Dendrocalamus parviflorus* Hack	Philippines
199.	*Dendrocalamus pendulus*	Malaya, Singapore.
200.	*Dendrocalamus sikkimensis* Gamble.	India
201.	*Dendrocalamus sinuatus.*	Malaya
202.	*Dendrocalamus siriceus* Munro	Thailand
203.	*Dendrocalamus strictus* Nees.	India, China, Thailand, Malaya, Sri Lanka, Indonesia.
204.	*Dendrocalamus strictus* Var. argentea	India
205.	*Dendrocalamus strictus* Var. *prainiana*	India
206.	*Dendrocalamus strictus* Var. *sericea*	India
207.	*Dinochloa aguilarii* Gamble	Philippines.
208.	*Dinochloa andamanica*	India
209.	*Dinochloa ciliata* Kurz	Philippines
210.	*Dinochloa compactiflora*	India, Burma.
211.	*Dinochloa distans*	Burma
212.	*Dinochloa elemeri* Gamble	Philippines.
213.	*Dinochloa luconiae* (Munro) Mer.	Philippines.
214.	*Dinochloa maclelandii* Kurz	India, Burma, Thailand.
215.	*Dinochloa orenuda* McClure	China
216.	*Dinochloa pubiranea* (Merr) Gamble.	Philippines
217.	*Dinochloa scandens* O.Kuntze	Malaya, Philippines, Indonesia.
218.	*Dinochloa scandens* Var. *angustifolia Hackel exMerr.*	Philippines
219.	*Dinochloa tjankorreh*	India
220.	*Dinochloa utilis McClure*	China
221.	*Drepanostachyum intermedium*	Nepal
222.	*Drepanostachyum khasianum*	Nepal
223.	*Fargesia ampullaris yi*	China
224.	*Fargesia chungii* (Keng) Wang ex yi	China
225.	*Fargesia edulis* yi	China
226.	*Fargesia farcata yi*	China
227.	*Fargesia grosa yi*	China
228.	*Fargesia setosa* yi	China
229.	*Fargesia spathacea* Franchel	China

230.	*Ferrocalamus stricutus* Hseuh ex yi	China
231.	*Gelidocalamus fangianus* (A. camus) Keng f. et Wen	China
232.	*Gelidocalamus kunishii* (Hayata) Keng f. et Wen	China
233.	*Gelidocalamus latifolius* Q H. Dai et T. chen.	China
234.	*Gelidocalamus rutilans* Wen	China
235.	*Gelidocalamus solidus* O. Dehou et C.S. Chao.	China
236.	*Gelidocalamus stellatus* Wen.	China
237.	*Gelidocalamus tesellatus* Wen	China
238.	*Gigantochloa albociliata* Munro	Thailand
239.	*Gigantochloa apus* Kurz	Bangladesh, Malaya, Indonesia.
240.	*Gigantochloa aspera* Kurz	Philippines, Java.
241.	*Gigantochloa atter* Munro	India, Malaya, Indonesia.
242.	*Gigantochloa aff atter* (Hassk) Kurz ex Munro.	Malaya, Africa
243.	*Gigantochloa compressa*	Thailand
244.	*Gigantochloa hasskarliana* Back ex Heyne	Malaya, Thailand, Indonesia.
245.	*Gigantochloa kurzii* Gamble	Indonesia
246.	*Gigantochloa latifolia*	Malaya
247.	*Gigantochloa levis* (Blanco) Munro.	Philippines, Malaya, Singapore.
248.	*Gigantochloa ligulata* Gamble.	Malaya, Thailand..
249.	*Gigantochloa macrostachya* Kurz.	Thailand, India
250.	*Gigantochloa maxima*	Indonesia
251.	*Gigantochloa maxima* Var. viridis	Malaya
252.	*Gigantochloa maxima* Var. *minor.*	Malaya
253.	*Gigantochloa maxima* Var. *ridley.*	Malaya
254.	*Gigantochloa naname*	Singapore
255.	*Gigantochloa nigrociliata* Munro.	Thailand, Indonesia.
256.	*Gigantochloa pseudoarundinacea* (steud) Widjaja	Indonesia, Malaya.
257.	*Gigantochloa ridleyi*	Malaya, Singapore.
258.	*Gigantochloa robusta*	Indonesia
259.	*Gigantochloa scortechnii*	Malaya
260.	*Gigantochloa seribneriana* Merril	Philippines
261.	*Gigantochloa tekserah*	Indonesia, India
262.	*Gigantochloa verticillata* Munro.	India, Malaya, Indonesia.

263.	*Gigantochloa wrayi*	Malaya
264.	*Guadua amplexifolia*	Venezuela, Nicaragua, Honduras.
265.	*Guadua angustifolia*	Columbia, Equador. Costa Rica
266.	*Guadua capitaya*	Costa Rica
267.	*Guadua chacoensis*	Costa Rica
268.	*Guadua inermis*	Mexico
269.	*Guadua paraguayana*	Paraguay
270.	*Guadua philippinensis*	Philippines
271.	*Guadua superba*	Brazil
272.	*Indocalamus latifolius* (McClure)	China
273.	*Indocalamus longiauricus* Handel-Mazz	China
274.	*Indocalamus mogoi* (Nakai) Keng. f	China
275.	*Indocalamus tessellatus* (Munro) Keng. f	China
276.	*Indocalamus wightiana*	India
277.	*Indocalamus walkerianus*	India
278.	*Indosasa crassiflora* McClure	China
279.	*Indosasa glabrata* Chu ex Chao	China
280.	*Indosasa shibatacaoides* McClure.	China
281.	*Indosasa sinica* chu & Chao.	China
282.	*Leleba floribunda* Nakai	Philippines
283.	*Leleba ventricosa* McClure	Philippines
284.	*Lingnania chungii* McClure	China
285.	*Lingnania wenchouensis* Wen.	China
286.	*Melocalamus compactiflorus* Benth.	Thailand
287.	*Melocanna baccifera* (Roxb) Munro.	India, Burma
288.	*Melocanna bambusoides* Trin.	India
289.	*Nastus elatus*	Indonesia
290.	*Nastus elegantissimus*	India, Indonesia
291.	*Nastus hooglandii*	Malaya, Indonesia
292.	*Nastus holtumianus*	Malaya
293.	*Nastus longispicula*	Malaya
294.	*Nastus obtusus*	Malaya
295.	*Nastus productus*	Malaya, Thailand
296.	*Nastus rudimentifer*	Malaya
297.	*Nastus reholttumianus*	Malaya
298.	*Nastus schechter*	Philippines
299.	*Nastus schmudtzu*	China
300.	*Neohouzeoua dulloa*	Burma, Bangladesh. India
301.	*Neohouzeoua helferi*	India, Burma.
302.	*Neohouzeoua mekongensis*	Indochina
303.	*Neohouzeoua stricta*	Burma
304.	*Neosinocalamus affinis* (Rindle) Keng f.	China

305.	*Neosinocalamus beecheyanus* (Munro) Keng f & Wen. ined.	China
306.	*Neosincalamus distegius* (Keng & Kenf f.) Keng f & Wen. ined.	China
307.	*Ochlandra beddomei* Gamble	India
308.	*Ochlandra brandisii* Gamble	India
309.	*Ochlandra ebracteata* Raizadda & Chaterjee	India
310.	*Ochlandra hosseusii* Pilger	Thailand
311.	*Ochlandra rheedii* Benth	India
312.	*Ochlandra scriptoria* (Denst) C.E.C. Fisher	India
313.	*Ochlandra setigera* Gamble	India
314.	*Ochlandra sivageriana* Camus	India
315.	*Ochlandra stridula* Thwaites	Sri Lanka
316.	*Ochlandra stridula* Var. *maculata*.	Sri Lanka
317.	*Ochlandra talboti*	India
318.	*Ochlandra travancorica* Benth	India
319.	*Ochlandra wightii*	India
320.	*Oligostachyum sulcatum* Wang ex ye.	China
321.	*Oxytenanthera abyssinica* (A Rich) Munro	Kenya
322.	*Oxytenanthera albociliata* Munro	India
323.	*Oxytenanthera bourdellonii* Gamble.	India
324.	*Oxytenanthera fetix*	China
325.	*Oxytenanthera hosseusii* Pilger	China
326.	*Oxytenanthera monadelpha*	Sri Lanka
327.	*Oxytenanthera monostigma* Bedd.	India
328.	*Oxytenanthera nigrociliata* Munro.	India
329.	*Oxytenanthera parvifolia* Brandis.	Burma
330.	*Oxytenanthera ritcheyi*	India
331.	*Oxytenanthera stocksii* Munro	India
332.	*Oxytenanthera thwaitesii* Munro.	India
333.	*Phyllostachys angusta* McClure	China
334.	*Phyllostachys aurea* Carr	China, Japan, Indonesia
335.	*Phyllostachys aurea* Var. *pekinensis* J.L. Luf Nov	China
336.	*Phyllostachys aurea* Var. *spectabilis.* (Chu chao) Lu	China
337.	*Phyllostachys aureosulcala* McClure.	China
338.	*Phyllostachys assamica*	India
339.	*Phyllostachys bambusoides* Sieb.	India, China, Japan, Philippines.
340.	*Phyllostachys bambusoides* Var. *aurea.* Makino.	Philippines
341.	*Phyllostachys decora* McClure	Japan
342.	*Phyllostachys edulis* Makino	Japan, Philippines.
343.	*Phyllostachys flexuosa* (Carr) Aet. C. Rly.	Japan, China

344.	*Phyllostachys glauca* McClure	Japan
345.	*Phyllostachys henonis*	Japan
346.	*Phyllostachys heterocycla* (Carr) Matsumi.	Japan
347.	*Phyllostachys heteroclada* oliv	China
348.	*Phyllostachys iridiscens*	Malaya
349.	*Phyllostachys manii*	India
350.	*Phyllostachys meyeri* McClure	China
351.	*Phyllostachys nigra* Munro	Philippines, Indonesia.
352.	*Phyllostachys nigra* Var *henonis* (Mitf Stapf ex Rendle.	Japan, China, Philippines, Indonesia.
353.	*Phyllostachys nidularia* Munro	Japan
354.	*Phyllostachys nuda* McClure	Japan, China
355.	*Phyllostachys praecox* Chou ex Chao.	China
356..	*Phyllostachys propinqua* McClure	China
357.	*Phylostachys purpurata* McClure C.V.	China
358.	*Phyllostachys pubescens* Mazel ex H Lehare.	China
359.	*Phyllostachys rubromarginata* McClure	China
360.	*Phyllostachys reticulata* C Koch.	China
361.	*Phyllostachys sedan*	Burma
362.	*Phyllostachys stimulosa* zhou et lin.	China
363.	*Phyllostachys viridis* (young)McClure.	Japan.
364.	*Phyllostachys vivax* McClure	Japan
365.	*Phyllostachys vivax* Var. Huang wenzhu L.U.tonov.	China
366.	*Pleiblastus amarus* (Keng) Keng f.	Japan
367.	*Pleioblastus chino* Makino	China
368.	*Pleioblastus communis*	China
369.	*Pleioblastus fortunei*	China
370.	*Pleioblastus gramineus* (Bean) Nakai.	China
371.	*Pleiobastus hindsii*	China
372.	*Pleioblastus hisiaenchuensis* Hsiung	China
373.	*Pleioblastus maculatus* McClure	China
374.	*Pleioblastus matsunoi* (Makino) Nakai.	China
375.	*Pleioblastus oleosus* Wen	China
376.	*Pleioblastus ovatoauritus* Wen, ined.	China
377.	*Pleioblastus pygmaeus*	China
378.	*Pleioblastus simonii* (Carr) Nakai.	Japan, China
379.	*Pseudoxytenanthera* monadelpha	Sri Lanka
380.	*Pseudosasa amabilis* (McClure) Keng g.	China
381.	*Pseudosasa contori* (Munro) Keng f.	China
382.	*Pseudosasa japonica.*	China, Japan
383.	*Pseudosasa longiligula* Wen.	China

384.	*Pseudosasa notata* Wang ex ye.	China
385.	*Pseudosasa orthotropa* Chen ex Wen.	China
386.	*Pseudostachyum polymorphum*	India
387.	*Racemobambos ceram*	Malaya
388.	*Racemobambos congesta*	Malaya
389.	*Racemobambos gibbsiae*	Malaya
390.	*Racemobambos glabra*	Malaya
391.	*Racemobambos hirsuta*	Malaya
392.	*Racemobambos hirta*	Malaya
393.	*Racemobambos multiramose*	Malaya
394.	*Racemobambos rigidifolia*	Malaya
395.	*Racemobambos schultzei*	Malaya
396.	*Racemobambos setifera*	Malaya
397.	*Racemobambos tesselata*	Malaya
398.	*Schizostachyum aciculare*	Malaya, Thailand
399.	*Schizostachyum acutiflorum* Munro.	Philippines
400.	*Schizostachyum alopecuras*	New Guinea
401.	*Schizostachyum biflorum*	Indonesia
402.	*Schizostachyum blumei*	Indonesia
403.	*Schizostachyum blumeana*	India
404.	*Schizostachyum brachycladum* Kurz	Malaya, Indonesia
405.	*Schizostachyum brachythyrsus*	New Guinea
406.	*Schizostachyum caudatum*	Indonesia
407.	*Schizostachyum curranii* Gamble	Philippines
408.	*Schizostachyum dielsianum* (Pilger) Merr.	Philippines
409.	*Schizostachyum diffusum* (Blanco) Merr.	Philippines
410.	*Schizostachyum fenixii* Gamble	Philippines
411.	*Schizostachyum funghomii* McClure	China
412.	*Schizostachyum gracile*	Malaya
413.	*Schizostachyum grande*	Malaya
414.	*Schizostachyum hallieri* Gamble	Malaya
415.	*Schizostachyum hainanensis* Merr.	China
416.	*Schizostachyum hirtiflorum* Hack	Philippines
417.	*Schizostachyum iraten*	Indonesia
418.	*Schizostachyum insulare* Ridley	Malaya, Thailand
419.	*Schizostachyum jaculans*	Malaya
420.	*Schizostachyum lima* (Blanco) Merr.	Philippines, India
421.	*Schizostachyum lumampao* (Blanco) Merr.	Philippines
422.	*Schizostachyum longispiculatum* Kurz.	Malaya, Thailand
423.	*Schizostachyum luzonicum* Gamble	Malaya
424.	*Schizostachyum merrillii* Gamble	Philippines
425.	*Schizostachyum mucronatum* Hack	Philippines
426.	*Schizostachyum palawanese* Gamble	Philippines
427.	*Schizostachyum pseudolima* McClure	China
428.	*Schizostachyum rogersii*	Burma, India
429.	*Schizostachyum serpentinum*	Indonesia

430.	*Schizostachyum terminale*	Malaya
431.	*Schizostachyum textorium* (Blanco) Merr.	Philippines
432.	*Schizostachyum toppingii* Gamble	Malaya
433.	*Schizostachyum whitei*	New Guinea
434.	*Schizostachyum xinwuensis* Wen	China
435.	*Schizostachyum zollingeri* steud	Malaya, Indonesia
436.	*Sasa Kazsa*	Japan
437.	*Sasa kurilensis* (Ruprecht)	Japan
438.	*Sasa palmata* Makino	Japan
439.	*Sasa paniculata*	Japan
440.	*Sasa purpurascens*	Japan
441.	*Sasa quingyuanensis* Hu.	China
442.	*Sasa sinica* Keng	China
443.	*Semiarundinaria lubricata* Wen	China
444.	*Semiarundinaria nitida*	China
445.	*Semiarundinaria patlingii*	India.
446.	*Semiarundinaria yashadake*	Japan
447.	*Sinobambusa anaurita* Wen	China
448.	*Sinobambusa edulis* Wen	China
449.	*Sinobambusa giganteus* Wen	China
450.	*Sinobambusa glabrescens* Wen	China
451.	*Sinobambusa intermedia* McClure	China
452.	*Sinobambusa latiflorus*	China
453.	*Sinobambusa nephroaurita* chu ex Chao.	China
454.	*Sinobambusa rubroligula* McClure	China
455.	*Sinobambusa tootsik* Makino	China
456.	*Sinobambusa tootsik* var *leata* (McClure) Wen.	China
457.	*Sinocalamus affinis* (Rendle) McClure.	China
458.	*Sinocalamus biciatricatus*	China
459.	*Sinocalamus latiflorus*	China
460.	*Sinocalamus oldhami*	China
461.	*Shibatea kumasaca* (zollinger) Makino.	Japan
462.	*Thamnocalamus aristatus*	India
463.	*Thamnocalamus falconeri*	India
464.	*Thamnocalamus prainii*	India
465.	*Thamnocalamus spathaceus*	China
466.	*Thamnocalamus spathiflorus*	Nepal, India
467.	*Teinostachyum attenuatum*	Srilanka
468.	*Teinostachyum beddomei*	India
469.	*Teinostachyum dullooa*	India
470.	*Teinostachyum griffithii* Munro	Thailand, Burma, India
471.	*Teinostachyum helferi*	Burma
472.	*Teinostachyum wightii*	India
473.	*Tetragonocalamus agulatus*	China

474.	*Thyrsostachys oliveri*	India, Burma
475.	*Thyrsostachys siamensis*	Thailand
476.	*Yushania chungii* (Keng) Wang et ye.	China
477.	*Yushania confusa* (McClure) Wang et ye.	China
478.	*Yushania hasihirsuta* (McClure) Wang & ye.	China
479.	*Yushania hirticaulis* Wang ex ye.	China
480.	*Yushania keng*	Philippines
481.	*Yushania niitakayamensis* (Hayata) Keng f.	Philippines
482.	*Yushania wixiensis* yi	China

The species listed in table 10 are by no means complete. It only contains species commonly known, and used for various purposes such as constructions, agricultural implements, and articles of every day use such as furniture, tool handles, utility articles, ladders, scaffolding and industrial raw materials for paper making, rayon pulp, veneers, and decorative articles etc., as well as for environmental control eg control of soil erosion, and aesthetics.

BAMBOOS NOT YET SCIENTFICALLY DESCRIBED

Research on bamboos so far has centered around utility. As such a number of species not yet scientifically described and named, and only known by local names are also being used. Examples of such species in India are 'boom' from Tripura, 'bushy dwarf' form Shillong, 'hard jati' and 'khupri' from Nagaland, 'murli' from Assam, 'paura' from Tripura, 'tachur' and 'tapin' from west Siang in Arunachal Pradesh, and 'bee' from Lower Sabansari from Arunachal Pradesh.

In Sudan, bamboo musical instruments are made exclusively of a bamboo known locally as 'bambu hitam'. It has been tentatively referred to as *Gigantochloa aff. atter* (Hassk) Kurz ex Munro. The leaf anatomy of this species is very different from that of the true *Gigantochloa atter* (Hassk) Kurz ex Munro which is known locally as 'bambu temen'. As floral characters of both are unknown, their taxonomic status has not been satisfactorily decided.

In China also there are some bamoos known only by their local names such as 'cheng ma quing', 'quing ma cheng', 'guabgxi' etc. In Nepal there are some bamboos suspected to be some species of *Drepanostachyum* known as 'malingo nigalo' some of which are used as edible shoots, and some as superior quality weaving material. Further, in Nepal, there are also some bamboos suspected to be species of *Thamnocalamus* known as 'goonre nigalo', 'chigar' and 'jaributo', all of which are an important food source for black bear and impeyan pheasant.

There is also some confusion in the revision of the genus *Arundinaria* by Nakai. In 1966 Dr. McClure of Smithsonian Institute U.S.A., quoted Hiroshi Usuii's 1957 report in "The Bamboos", and pointed out that *Pleioblastus* is the same as *Arundinaria*. Nakai who did much work on bamboos was also perplexed about *Pleioblastus* and later proposed a new genus called *Nipponocalamus*. But several of *Pleioblastus* species described by Nakai has been shown to belong to *Arundinaria* by Makino.

REQUIREMENTS FOR A SYSTEMATIC INVESTIGATION OF BAMBOOS

A systematic investigation of bamboos for the purpose of classification and taxonomical studies would entail survey and field studies and collection of material along with ecological data, and establishment of experimental plantations and maintenance of living collections in a number of different places, as well as maintenance of a herbarium for flowering parts and vegetative parts, also a xylarium for woody culms, and fruit and seed collections, in each locality where living collection are maintained. This will ensure that flowers, fruits, seeds, leaves, culm sheaths, and culms, taken up for investigations are from the same species, and correlation of data from morphological, anatomical, cytological and biochemical analysis will yield more reliable results. Moreover such studies carried out in different places will yield data on geographical differentiation between taxa caused by spatial separation, and major patterns of evolution, and pathways of migration as well as the ways in which a taxon has adapted to the different environments which it has encountered during its migrations.

In India, in the last decade an effort has been made in this direction by the National Bureau of Plant Genetic Resources, New Delhi, and its stations in Trichur, Shillong, and Ranchi, as well as by the ICAR Research Complex for hill regions. The Arunachal Pradesh centre has initiated the collection and build up of genetic diversity of bamboos for evaluation. Germplasm collections have also been established at New Delhi, Trichur, and Ranchi centres and studies have been carried out on bamboo germ plasm under the All India Coordinated Project at Arunachal Pradesh centre located at Basar from 1984 onwards.

In Thailand, bamboo collections have been set up in three regions of the country viz. (i) at Mace Sa Botanical Gardens in Chiangmai in the north; (ii) at Hin lap Research Station in Kanchanabir in central Thailand; (iii) at Songkla Forest Research Station in Songkla in south Thailand. The living collections at Kanchanbur and Songkla are on flat land, and the one in Chiangmai, is in a hilly area. There are also living bamboo collections at Dong Lam Seed Orchard Khonkhaen province, and Thung Salang Luang in Phiranulok province of Thailand.

In Sri Lanka, the International Development Research Centre (IDRC) has since 1984 supported a project which aims at determining the distribution and availability of individual species. Sri Lanka also has a collection of bamboos in three botanic gardens, the best collections are in Paradeniya Botanic Gardens.

In Nepal the Forest Research Project (FRP) which is a part of the Forest Survey and Research Centre has a section entirely on bamboos and they carry out research on identification and distributions of bamboos.

In Africa a number of bamboo species are being cultivated in (i) Amani Arboretum in north Tanzania (ii) Entebbe Botanic Gardens in south Uganda; (iii) Kuguga Arboretum in central Kenya; and (iv) Zomba Arboretum in Malawi.

There are also living collections of bamboos in China, Japan, Philippines and Indonesia.

REFERENCES

Abbot, H.C.de S (1886) - Certain chemical constituents of plants considered in relation to their morphology and evolution. *Bot. Gaz.***11** (pp. 270 - 272).

Abston, R.E. and Turner, B.L. (1959) - Application of paper chromatography to systematics. Recombination of parental biochemical components in a Baptista hybrid population. *Nature*. London. **184.** (pp. 285 -286).

Ambler, R.P. (1972) - Sequence data acquisition for the study of phylogeny. In Recent developments in the chemical study of protein structure edited by A.Previcro, J.F. Peechere and M.A Colleti Previcro. (pp. 289 - 305).

Antonovics. J. and Bradshaw, A.D. (1970) - Evolution in closely adjacent plant populations, VII clinal pattern at a mine boundary. *Heredity*. **25**, (pp. 349 - 362).

Arber, A. (1926) - The flowers of certain Bambuseae. Annals of Botany Vol. XL no. CL VIII, April (pp. 447 - 469).

Arber, A. (1927) - Studies in the Gramineae II. Abnormalities in *Cephalostachyum variegatum* Kurz and their bearing on the interpretation of the bamboo flower. Annals of Botany 41.

Arber, A. (1930) - Studies in the Gramineae IX (i) The nodal plexus, (ii) Amphivasal bundles. Annals of Botany Vol. CLVI London. (pp. 593 - 620).

Arber, A. (1934) - The Gramineae - a study of cereal bamboo and grass. Cambridge University Press, Cambridge, England.

Asano, S. et al (1979) - Taxonomic and ecological observations of *Sasa borealis* Makino. Rep. of Fiji bamoo garden.

Bahadur, K.N. and Naithani, H.B. (1978) - On a rare Himalayan bamboo. *Ind. Jour. For.* Vol. **1**, No. 1. (pp. 39 - 45).

Bahadur, K.N (1979) - Taxonomy of bamboos. *Ind. Jour. For.* Vol. **2**, No. 3 (pp. 222 - 241).

Bate - Smith E.C. (1973) - Chemotaxonomy of Germanium *Bot. Jor. Linn. Soc.* **67.** (pp. 347 - 359).

Bendich A.J. and Bolton, E.T. (1967) - Relatedness among plants as measured by the DNA - agar techinque. *Plant physiology*. Lancaster. **42.** (pp. 959 - 967).

Bennet, S.S.R. and Gaur R.C. (1990) - Nomenclature of Burmese bamboos *Melocanna humilis* Kurz. Indian Forester Vol. **116** No. 8, (pp 648 - 649).

Bennet, S.S.R. and Gaur R.C. (1990) - Thirty seven bamboos growing in India Dehradun Forest Reserach Institute. (100 Pages)

Bentham G. (1983) - Gramineae. In Genera plantarum by Bentham, G and Hooker, J.D.Reeve London England. Vol. **3**, (pp. 1207 - 1215).

Bissen.S.S, Mishra, G.P. and Sharma, S.C. (1988) - Scanning electron microscopic studies on culm and leaf epidermis of Indian bamboos. *Ind. For.* Vol. **114**. No. 10 (pp. 656 - 669).

Blatter, E. and Parker, R.N (1929) - Indian bamboos brought upto date *Ind. For.* Vol. **55** (pp. 541 — 562).

Brady, J. (1979) - Biological clocks. Edward Arnold Ltd. Camelot Press. Southampton, Britain.

Brandis, D. (1899) - Biological notes on Indian Bamboos. *Ind. For.* Vol. **25** (pp. 1-25).

Brown, W.H. and Fischer, A.F. (1918) - Philippine Bamboos, Bulletin No. 15. Bureau of Printing. Manilla, Philippines.

Camus,E.G. (1913) - Les Bambusees: monographic, biologic, culture, principaux usages. Lechevalier Paris, France. (pp. 1-215).

Collins, R.C. (1957) - Taxonomy of higher plants. *Amer. Jour. Bot.* Vol **44.** (pp. 188 - 196).

Colless, D.H. (1967) - An examination of certain concepts in phenetic taxonomy. *Syst. zool.* Vol. **16** (pp. 6-27).

Crowson, R.A. (1972) - A systematists look at cytochrome C. Jor; *Molec. Evol.* Vol **2.** (pp. 28-37).

Danielsson, C.E. (1949) - Seed globulins of the Gramineae and Leguminoseae. *Biochem. Jour.* **44** (pp. 287 - 400).

Davis, P.H. and Heywood V.H (1963) - Principles of Angiosperm Taxonomy. Oliver and Boyd, Edinborough and London.

Eglinton, G., Gonzales, A.G., Hamilton, R.J. and Raphael R.H. (1962) - Hydrocarbon constituents of the wax coatings of plant leaves — A taxonomic survey. *Phytochemistry*, (pp. 89-120).

Fairbrother, D.E. and Johnson, M.A. (1961) - The precipitin reaction as an indicator of relationships in some grasses. *Recent Advances in Botany* (pp. 116-120). Univ. of Toronto Press.

Fairbrother, D.E. (1968) - Chemosystematics with emphasis on systematic serology. In Modern Methods in Plant Taxonomy edited by V.H. Heywood, Academic Press, London.

Fujimoto,Y. (1966) - Classification of Bambusoideae based on leaf characters especially ligule portion. Rep. of Fuji Bamboo Gard. 11.

Gamble, J.S. (1896) - The Bambuseae of British India. Ann. of the Royal Bot. Gard. Calcutta.

Gamble, J.S. (1910) - The Bamboos of the Philippine Islands. Phil. Jour. of Science (Botany) Vol. **5**, No. 4 (pp. 267-281).

Ghosh S.S. and Negi, B.S. (1960) - Anatomy of Indian Bamboos Part 1. Epidermal features of *Bambusa arundinacea, B.polymorpha, B.vulgaris, Dendrocalamus membranaceous, D.strictus,* and *Melocanna bambusoides. Ind. For.* Vol. **86** No. 12 (pp. 719-727).

Gibbis, R.D. (1963) - History of chemical Taxonomy in Chemical Plant Taxonomy, edited by T.swain (pp. 41 - 88) Academic Press, London and New York.

Gregory, W.C (1941) - Phylogenetic and cytological studies in Ranunculaceae. *Trans. Am. Phil. Soc.* N.S. **31** (pp. 443-521).

Grosser, D. and Liese, W (1971) - On the Anatomy of Asian bamboos with special reference to their vascular bundles. Wood Science and Technology Vol. **5** (pp. 290-312).

Grosser, D. and Liese, W (1973) - Present status and problems of bamboo classification Jour. of Arnold Arboretum. Vol. **54** No.2 (pp. 293-308).

Grosser, D. and Zamuco, G.I.Jr. (1971) - Anatomy of some Bamboo species in the Philippines. *Phil. Jour. of Science* Vol. **100** No 1 (pp. 57-73).

Gupta, B.N and Pattanath P.G. (1976) - Variation in stored nutrient in culms of *Dendrocalamus strictus* and their effect on rooting of culm cuttings. Indian Forester. Vol. **102**, No. 4

Haryonto Yudodibroto (1985) - Bamboo research in Indonesia. Proc. of Int. Bamboo workshop Hangzhou. Peoples Republic of China, October 6th - 14th.

Heywood, V.H. and McNeill,J. (1964) - Phenetic and phylogenetic classifications, Systematic Association, London.

Heywood, V.H. (1967) - Plant Taxonomy, Arnold. London.

Heywood, V.H. (1973) - Taxonomy and Ecology. Academic Press, London.

Hildebrand, F.H. (1954) - Antekeningen over Javanese Bambus Soorten. Notes on Javanese bamboo species. Indonesian Forest Research Institute. Report 66. (82).

Hiroshi Usuii (1985) - Morphological studies on prophylls and their systematic significance. Proc. of Int. Bamboo workshop, Hangzhu. Peoples Republic of China, October 6th to 14th.

Holttum, R.E. (1946) - Classification of Malayan Bamboos. *Jour. of Arnold Arboretum.* Vol **27** (pp. 340 - 346).

Holttum, R.E. (1956) - Classification of Bamboos Phytomorphology Vol. **6** (pp. 73-90).

Holttum, R.E. (1958) - Bamboos of the Malaya Peninsula. Bulletin of Botanic gardens, Singapore. No. 16.

Holttum, R.E. (1967) - Bamboos of New Guinea Kew Bulletin Vol **21.** No. 2 (pp. 263-292).

Hooker, J.D. (1854) - Bamboos: *Chusquea abietifolia* Himalayan Journal Vol 1.

Hubbard. C.E. (1948) - Gramineae: In the British flowering plants by Hutchinson. (pp. 284-348) Gawthorn, London.

Hutchinson, J. (1959) - The Families of Flowering Plants: Monocotyledons. Second edition Vol II. Oxford Clarendon Press, London.

Jain, S.K. (1969) - Comparative ecogenetics of two Avena species occurring in central California. *Evolutionary Biology.* Vol. **3**. (pp. 73-113).

Jain, S.K. and Bradshaw, A.D. (1966) - Evolutionary divergence among adjacent populations. I. The evidence and its theoretical analysis. *Heredity. Vol* **21**. (pp. 407-441).

Janzen, D.H. (1976) - Why bamboos wait so long to flower. *Ann. Rev. of Ecol. Syst.* Vol. **7** (pp. 347-391).

Jeffrey, C. (1981) - Introduction to Plant Taxonomy Cambridge University Press, London.

Jian Xin and LiQion (1985) - Observations on vascular bundles of bamboos native to China Proc. of Int. Bamboo workshop. October 6th - 14th (pp. 227-229).

Johnson, B.L. and Thein, M.M. (1970)-Assessment of evolutionary affinities in Gossypium by protein electrophoresis. *Amer. Jour, Bot,* Vol. **57** (pp. 1081-1092).

Johnson, B.L. and Hall, O. (1965) - Analysis of phylogenetic affinities in Triticinae by protein electrophoresis. *Amer. Jour. Bot.* Vol. **52**. (pp. 506-513).

Karelstchicoffs (1868) - Die falten formigen verdickungen in den zellen einiger Gramineen. Bulletin de la societe Imperiale des Naturalistes de Moscou XII, No. 7. (pp. 180-190).

Kellman, M.C. (1980) - Plant Geography. Published by Methuen. London and New York.

Kitamura, H. Sakamoto, I. Nagashima, T. (1974) - On the quadrangular bamboo culms - forms of the culms and the anatomy with special reference to fibres and vascular bundles. Bulletin utsunomiya University Forests. Vol. **11**. (pp. 71-86).

Lalita Kumari, Reddy, P.R. and Jagdish, C.A. (1985) - Identification of species of Bambusa by electrophoretic pattern of peroxidases. *Indian Forester* **111** No. 8. (pp 603-609).

Lantican, C.B. Palijjon, A.M. and Saludo, C.G. (1985) - Bamboo research in the Philippines .Proc. of Int. Bamboo workshop. Hangzhou, peoples Republic of China, October 6th - 14th.

Lindayen, T.M. Valbuena, R.R. and Tomoland, F.N. (1969) - Erect bamboo species in the Philippines. *Philippine Lumberman.* Vol **15**. No. 1.

Madan Gopal, P.G. Pattanath and R.C. Thapaliyal (1972) - Exploratory trial of propagation of *Dendrocalamus strictus* and *Bambusa polymorpha* by culm cuttings. Proceedings and Technical papers of symposium on Man Made Forests in India. Dehradun 1972, organized by Forest Research Institute, Dehradun.

Marshal, D.R. and Jain, S.K. (1968) - Phenotypic plasticity of *Avena fatura* and *A. barvata Amer. Nat.* Vol **102.** (pp. 457-467).

McClure, F.A. (1934) - Inflorescene in *Schizostachyum* Nees. Jour. of Washington Acad. of Science, Vol **24.** (pp. 541-548).

McClure, F.A. (1963) - A new feature in bamboo rhizome anatomy. Rhodora Vol. **65** (pp. 63-65).

McClure, F.A. (1973) - Genera of bamboos native to the new world. In Gramineae: Bambusoideae edited by T.R. Soderstrom. Smithsonian contributions to Botany no. 9 (148 pages).

Merrill, E.D. (1905) - A review of the species described in Blanco's Flore de Filipinas. Philippines Government Laboratory Publication, XXVII 93.

Metcalfe, C.R. (1954) - An Anatomists view on angiosperm classification. Kew Bulletin No. 6 (pp. 427-440).

Metcalfe, C.R. (1960) - Anatomy of the monocotyledons I. Gramineae: bamboos, (pp. 578-585), Oxford University Press, London.

Michener, C.D. (1970) - Diverse approaches to systematics. Evolutionary Biology, Vol. **4.** (pp. 1—37).

Moore, D.M. (1968) - The karyotype in taxonomy. In modern methods in plant taxonomy edited by V.H. Heywood. Academic Press, London.

Munro, W. (1868) - Monograph of the Bambusaceae, Transactions of the Linean Society of London. *Botany* Vol. **26** (pp. 1-15).

Murata, G. (1979) - Taxonomical notes no. 13. Acta Phytotaxonomica Geobotanica. Vol **30.**

Naithani, H.B. (1990) - Nomenclature of Indian species of *Oxytenanthera* Munro. Journal of the Bombay Natural History Society 87, (pp. 439-440).

Naithani H.B. (1990) - Two new combinations of bamboos. *Indian Forester* **116** No 12. (pp. 990 - 991).

Naithani H.B. (1992) - A new species of bamboo *Schizostachyum* Nees from Arunachal Pradesh, India. *Indian Forester* **118** No. 4 (pp. 230 -231).

Naithani H.B. (1993) - Contributions to the taxonomic studies of Indian bamboos. 2 vols. Ph D. thesis - pp. 481.

Nakai, T. (1925) - Two new genera of bambusaceae with special remarks on the related species growing in eastern Asia: *Jour. Arnold Arboretum.* Vol **6.** (pp. 145-153).

Nakai, T. (1933-1935) - Bambuseae in Japan proper. Parts 1-6 *Jour. Jap. Bot.* Vol **9**, No. 11.

Nees., C.G. (1841) - Gramineae - Bamboos. Florae Agricae australioris Vol. **1** (pp 460-464).

Neela de zoysa et. al (1988) - Some aspects of bamboo and its utilization in Sri Lanka. Proc. Int. Bamboo workshop held in Cochin. Nov. 14th - 18th.

Nelson, A.P. (1965) - Taxonomic and evolutionary significance of lawn races of *Prunella vulgaris. Brittonica* Vol **17**. (pp.160-174).

Ng, F.S.P. and Shamshaddin (1980) - Bamboos of Malaysia. Proc of Bamboo workshop on Bamboo Research in Asia. Singapore 28th to 30th May.

Nicholoson, J.W. (1922) - Notes on the distribution and habitat of *Dendrocalamus strictus* and *Bambusa arundinacea* in Orissa. *Ind. For.* Vol **48**. (pp. 425-428).

Ohrenberger, D. and Goerrings, J. (1989) - Bamboos of the world. Published by R.P. Singh Gahlot on behalf of International Book Distributors P.o. Box 4, Dehradun, India.

Olmsted, C.E. (1944) - Growth and Development of range grasses. *Bot. Gaz.* Vol. **106** (pp. 46–74).

Ordinario, F.F. (1976) Distribution, present level of technology and development potentials of bamboos. Paper presented at the Philippine Forest Research Society Symposium. 29th September 1976. College Laguna, Philippines.

Orloci, L. (1968) - Information analysis in phytosociology partition, classification and prediction. *Jour. Theor. Biol.*, Vol **20**. (pp. 271-284).

Oye, R. (1980) - Bamboo Research in Japan - Country Report. Proc of workshop on Bamboo Research in Asia held in Singapore. May 28th to 30th.

Parodi, L.R. (1936) - Les Bambusees indigenas en la Mesopotamia Argentina Review Argentina Agrom. Vol **3**. (pp. 229-244).

Parthasarathy, N. (1946) - Chromosome number in Bambuseae. *Current Science*. Vol 15. (pp. 233-234).

Pattanath, P.G. (1965) - Studies on the anatomy of Indian bamboos. Final report on the research project R38 of Wood Anatomy Branch, Forest Research Institute, Dehradun.

Pattanath, P.G. and K.Ramesh Rao (1969) - Epidermal and Internodal structure of the culm as an aid to identification and classification of bamboos. In Recent Advances in the Anatomy of Tropical Seed Plants edited by K.A. Chowdhury. Hindustan Publishing Corporation, Delhi, India. (pp.179-196).

Pattanath, P.G. (1972) - Trend of variation in fibre length in bamboos. *Indian Forester* Vol **98**. No. 4. (pp. 241-243).

Poterfield, W.M. (1923) - A new feature in the vascular anatomy as displayed by bamboo-particularly the young leaf sheath. China Journal of Science and Arts. Vol I. (pp. 273 - 279).

Poterfield, W.M. (1937) - Histogenesis in the bamboo with special reference to the epidermis. *Torrey Bot. Club Bulletin* Vol **64**. (pp. 421-432).

Quastler, H. and Sherman F.G. (1959) - Experimental Cell Research Vol **17.** (pp. 420-438).

Qureshi I.M. and Deshmukh T. (1962) - Bamboos of India. (Memeographed Information leaflet) FRI Dehradun.

Raizada, M.B. (1948) - A little known Burmese bamboo - *Sinocalamus Copelandi* Indian Forester. Vol **74**. (pp. 7-10).

Raizada, M.B. and Chaterjee, R.N. (1963) - Cuim sheaths as an aid to identification of bamboos. *Ind. For.* Vol **89**. No. 11 (pp. 744-756).

Raizada, M.B. and Chaterjee, R.N. (1963) - New bamboo from south India. *Ind. For.* Vol. **89.** No. 5 (pp. 362-364).

Richaria, R.N. and Kotwal J.P. (1940) - Chromosome number in bamboo - *Dendrocalamus strictus.* Indian Journal of Agricultural Science Vol. **10**. Page 1033.

Ridley, H.N. (1907) - Materials for the Flora of the Malayan Peninsula: Monocotyledons Botanic Gardens Singapore 3. (pp. 182-197).

Rumphinus, G.E. (1750) - Herbarium ambionense. M. uytwerf. Amsterdam, Netherlands.

Ruprecht, F.J. (1839) - Bambuseas monographic. Exponit. Acad. St. Petersburg. (pp.1-7).

Saleh Mohd. Norand K, and Wong M. (1985) - The Bamboo resource in Malaysia. Strategies for development. Proc. of Int. Bamboo Workshop. Hangzhou. Peoples Republic of China, October 6th - 14th.

Sakomsak Ramyarangsi (1985) - Bamboo Research in Thailand. Proc. of Int. Bamboo Workshop. Hangzhou. Peoples Republic of China, October 6th - 14th.

Samapuddhi, K. (1959) - A preliminary study in the structure and some properties of some Thai bamboos. No. R. 30. Royal Forest Dept. Bangkok.

Sharma, Y.M.L. (1980) - Bamboos in the Asia Pacific Region. Proc. of Workshop on Bamboo Research in Asia held at Singapore 28th to 30th May.

Shingenmatsu, Y. (1958) - Analytical investigation of the stem form of the important species of bamboo. Bull. Fac Agric. University. Miyazaki No. 3. (pp. 125-135).

Shi Quantai ànd Chao Ching Ju (1980) - Bamboos in China. Proc. of workshop on bamboo research in Asia held at Singapore. 28th to 30th May.

Snaydon, R.W. (1973) - Ecology genetics and speciation. Academic Press. Inc. Ltd. (London).

Socjatmi Dransfield (1980) - Bamboo taxonomy in the Indo Malayan region. Proc. of Workshop on Bamboo research in Asia held in Singapore 28th - 30th May 1980. Edited by Gilles Lessard and Any Chouinard.

Solbrig, O.T. (1968) - Fertility, Sterility and the species problem. In Modern methods in plant taxonomy. Edited by V.H. Heywood. Academic Press, London.

Stace, C.A. (1989) - Plant Taxonomy and Systematics. Second Edition. Edward Arnold. Division of Hodder Stoughton, London, Melbourne, Auckland.

Stapf. O. (1904) - On the fruit of *Melocanna bambusoides*. Trin. an endospermless viviparous genus of Bambusaceae. Trans. Lin. Soc. London (Bot) Vol. **6** (pp. 401-425).

Stapleton C. (1989) - Bhutanese bamboos a simple key and background information on the genera, Tsenden Vol **1**. No. 1. (pp. 26 - 30).

Subramanian K.N. and P.E. Bedell (1990)-Development of bamboo potential—A taxonomic problem, National seminar on bamboos organized jointy by Karnataka Forest Department and Bamboo society of India, Bangalore, 20th to 23rd November 1990.

Stebbins, J.L. (1950) - Variation and evolution in plants, Columbia Univ. Press, New york.

Sunchengzhi and Xie Guoen (1985) - Fibre morphology and crystallinity of *Phyllostachys pubescens* with reference to age. In Recent Research on Bamboos - Proc of Int Bamboo Workshop. Hangazhu. People Republic of China.

Suzuki, S. (1978) - Inden to Japanese Bambusaceae Gakken, Tokyo, Japan.

Takagi, T. (1965) - Taxonomic studies of genus sasa by flower, Bamboo 4.

Takenouchi, y. (1931) Morphologische and entwicklungsmechanische untersuchungen bei japanischen bambus arten. Memorial College of Science, Kyoto Imperial University (Series B)6. (pp.109-160).

Takenouchi, Y (1932) - Form and anatomical structure of bamboo culm. Japanese Forestry Society Journal Vol. **14**. No. 24.

Thoday, J.M. (1953) - Components of Fitness. S.E.B. Symp. 7. (pp. 96 - 113).

Thoday, J.M. and Gibson, J.B. (1962) - Isolation by disruptive selection. *Nature* **193** (pp.1164-1166) London.

Trotter, H. (1922) - Development of bamboos from natural seedlings (Dendrocalamus) strictus). *Ind. For.*, Vol **48** (pp. 531-536).

Troup, R.S. (1921) - Sliviculture of Indian Trees. Oxford Clarendon Press.

Turresson, G. (1922) - The genotypic response of the plant species to the habitat. *Hereditas.* **3**. (pp. 211-350).

Turner, B.L. (1970) - Molecular approach to population problems at the infraspecific level. In Phytochemical Phylogeny edited by J.B. Harborne. Acad. Press, London.

Turrill, W.B. (1942) - Taxonomy and Phylogeny Bot. Rev. **8**. (pp. 247 - 270, 473 -532, 655 - 707).

Ueda, K. and Numata, M (1961) - Silvicultural and ecological studies of a natural bamboo forest in Japan. Bull. Kyoto. Univ. Forests 33.

Vavilov. N.I. (1951) - The origin, variation, Immunity and breeding of cultivated plants. Chronica Botanica Co. Waltham.

Venkatraman, T.S. (1937) - Sugarcane x bamboo hybrids. *Ind. Jour. Agri. Science.* **7**. (pp. 513-514).

Vivekanandan, R. (1985) - Bamboo Research in Sri Lanka. Proc. of Int. Bamboo Workshop. Hangzhou. People's Republic of China, October 6th - 14th.

Watson, E.V. (1943) - The dynamic approach to plant structure and its relation to modern Taxonomic Botany. *Biol. Rev.* (Gr. Britain) **18**. (pp. 65-77).

Wen Taihui and Chou Wen wei (1985) - A study on the anatomy of vascular bundles of bamboos from China. Proc. of Int. Bamboo. Workshop October 6th - 14th (pp. 230-243) Hangzhou. Peoples Republic of China.

Williams, W.T. and Dale, M.B. (1965) - Fundamental problems in numerical taxonomy. In Advances in Botanical Research. Edited by R.D. Preston. Vol **2**. Acad. Press, London.

Wu Bo and Ma Naixun (1985) - Bamboo Research in China. Proc. of Int. Bamboo Workshop. Hangzhou. Peoples Republic of China, October 6th - 14th.

Zamuco, G.I. Jr and Tongacan, A.L. (1973) - Anatomical structure of four erect bamboos of the Philippines. Philippine Lumberman Vol **19**. No. 10, (pp. 20-31).

Zeven, A.C. and Zhukovsky, P.M. (1975) - Dictionary of cultivated plants and their centre of diversity. Centre for Agric. Publishing and Documentation. Wageningen.

Zhang Guangzhu and Chen Fuqui (1985) - Studies on bamboo hybridisation. Proc. of Int. Bamboo Workshop. Hangzhou. Peoples Republic of China October 6th-14th.

Zhang Guangzhu (1985) - Studies on chromosome number of bamboo species with clump rhizomes. Proc. of Int. Bamboo Workshop Hangzhou. Peoples Republic of China. October 6th - 14th.

Zomber (1944) - The Common local bamboo, Nyasaland Quarterly Journal Vol **4**. No. 3 (pp. 8-13).

GLOSSARY

Adnate - united to a number of another series.

Adaxial - next to the axis.

Allopatry - The phenomenon relating to groups of plants that could interbreed but do not beacuse they are geographically separated.

Amino acid - The basic building block of proteins.

Amphoteric - Chemically reacting as acidic to strong bases and as basic towards strong acids.

Amphivasal bundle - A vascular bundle in which a central phloem is surrounded by xylem.

Antigen - Any substance that the body regards as foreign and consequently elicits an immune response. They are usually proteins.

Antibody - A protein produced in response to the entry into the body of a foreign substance termed as antigen in order to render it harmless. An antibody antigen reaction is highly specific. Specific antigens stimulate the formation of specific antibodies.

Apomixis - Reproduction which has superficial appearance of ordinary sexual cycle but occurs actually without fertilization.

Auricle - ear-like extensions of the blade portion of culm sheaths of bamboos.

Autoradiography - A process commonly used in tracer work, where an image is obtained by placing a thin biological specimen containing a radioactive isotope in contact with a photographic plate and exposing for a suitable period and developing. The image shows the distribution of radioactive element in the specimen.

Awn - Bristle-like appendage.

Biological Clock - The physiological processes that measure time by some means and do so in relation to environmental time cues such as day lengths, high tides, drought periods etc.

Bract - Modified leaf subtending the flower stalk or flower.

Caryopsis - a one seeded fruit with pericarp united to the seed.

Centromere - The constricted region of a nuclear chromosome to which the spindle fibres attach during division.

Chiasmata - Cross shaped structures commonly observed between sister chromatids during meiosis at the site of crossingover.

Chromatid - One of the two side by side replicas produced by chromosome division.

Cline - An increasing or decreasing gradient within a continuous population in the frequencies of different phenotypes or genotypes.

Collateral vascular bundle - A bundle in which phloem is on the same radius as the xylem and external to it.

Convergent evolution - The development of superficially similar structures in unrelated plants, usually because they live in the same kind of environment.

Culms - Stem of the bamboo.

Cybernetics - The theory of communication and control mechanisms in living things.

Deciduous - trees which are not evergreen.

Diploids - Individuals having two chromosome sets in their cells.

Distichous - arranged in two vertical ranks.

DNA - Deoxyribonucleic acid — a double chain of linked nucleotides having deoxyribose as their sugar. DNA is the fundamental substance of which genes are composed.

Ecoclinal variation - grades of variation within a species usually occurring as reactions to different ecological zones, within the species distribution.

Ecotype - A term used to mean a local race generally an ecological race with genotypes adapted to a particular restricted habitat as a result of natural selection within the local environments.

Ecological units - Combined physical, chemical, physiological, and biotic factors that is necessary for the survival of the species.

Edaphic subclimax - a relatively stable ecological community achieved at the end of a succession through factors relating to physical and chemical composition of the soil found in a particular area.

Electrophoresis - A technique used extensively in studying mixtures of proteins, nucleic acids, carbohydrates, enzymes etc. It is based on the movement of charged colloidal particles in an electric field and there are a large number of experimental methods.

Endemism - The process of being confined to a given region such as an island.

Enzymes - A large group of proteins produced by living cells, which act as catalysts in the chemical reactions upon which life depends. All enzymes are highly specific in their action and require precisely defined conditions of temperature and pH for their optimum performance.

Euploid - An individual composed of cells having a large number of complete chromosome sets.

Evolution - Cumulative changes in the characteristics of populations occurring in the course of successive generations related by descent.

Fascicles - A bunched tuft of branches.

Finger printing - The process by which characteristic spot patterns are produced by electrophoresis of the polypeptide fragments, obtained through denaturation of a protein with proteolytic enzymes.

Fractionation - Separation of a mixture into fractions of different properties by fractional distillation.

Geniculate - kneed, jointed or knotted

Genome - The entire complement of genetic material in chromosome set.

Genus - A term used in classification for a group consisting of a number of similar species.

Gene interchange - Movement of genes as a result of mating and gene exchange within populations.

Genetic code - The set of correspondences between nucleotide pair triplets in DNA, and amino acids in protein. It is the code by which inherited characteristics are handed down from generation to generation.

Genetic Manipulation - genetic engineering.

Germ plasm collection - A collection of genetic material of definite chemical and molecular constitution. Germ plasm is the physical basis of inherited qualities transmitted from generation to generation.

Glume - Chaffy bract, a pair of which occur at the base of bamboo spikelets enclosing them.

Herbarium - Collection of preserved plant specimens for identification and reference purposes.

Heterochromatin - Densly staining condensed chromosomal regions believed to be for the most part genetically inert.

Hexaploid - A polyploid with six chromosome sets in the somatic cells.

Inselberg - A steep sided eminence arising from a plain tract often found in semiarid regions of tropical countries.

Lemma - Outer bract of bamboo floret.

Ligule - Projection from inside junction of sheath and blade of bamboos.

Lodicule - A scale like perianth segment.

Monophyletic taxon - A group that is assumed to have originated from the same ancestor.

Nucleotide - An organic compound consisting of a nitrogen containing purine or pyrimidine base linked to a sugar (ribose or deoxyribose) and a phosphate group.

Outbreeding - Mating between unrelated or distantly related individuals of a species. Outbreeding populations usually show more variation than inbreeding ones and have a greater potential for adapting to environmental changes. Outbreeding increases the number of heterozygous individuals so that disadvantageous recessive characteristics tend to be masked by dominant alleles.

Paleas - Inner bract of the bamboo floret.

Panicle - Branched inflorescence.

Parenchyma - Tissue consisting of living thin walled isodiametric cells and permeated by a system of intercellular spaces containing air. In bamboo culms they form the ground tissue holding the fibrovascular bundles in place.

Peptide bonds - A bond joining two amino acids.

Peptide - Any group of organic compounds comprising two or more amino acids linked by peptide bonds. These bonds are formed by the reaction between adjacent carboxyl (—COOH) and amino (—NH2) groups with the elimination of water.

Phylogeny - The evolutionary history of an organism or group of related organisms.

Phytogeography-The Study of the distribution of world vegetation, with particular emphasis on the influence of the environmental factors that determine this distribution.

Phloem- A tissue that conducts food materials in vascular plants from regions where they are produced (notably the leaves) to regions such as the growing points.

Polyphyletic - Denoting a group of organisms, the members of which have originated from several different ancestors. Polyphyletic groups are not natural groups and do not have any place in natural classification.

Polyploid - A term used for a nucleus that contains more than two sets of chromosomes.

Polysaccharides - A group of carbohydrates consisting of long chains of monosaccharide (simple sugar) molecules.

Raceme - A type of racemose inflorescence in which the main flower stalk is elongated and bears stalked flowers.

Rachis - Axis.

Rachilla - Axis of bamboo spikelet.

Rank - The position or status of a taxon in classification heirarchy.

Reproductive Isolation - Isolation by various genetically controlled mechanisms which prevent gene exchange between two populations and preserve difference in the gene pools of populations previously achieved by natural selection and geographic isolation.

Rhizome - A horizontal underground stem which serves to propagate the plant vegetatively.

Sclereids - Also known as sclerenchyma is a tissue which gives mechanical support to plants and consists of thick walled cells usually lignified with a very small lumen. Two types of cells come under sclerenchyma - the fibres and the stone cells. The latter is known as sclereids.

Scutellum - Cotyledom of bamboo and other gramineae seed by which the embryo absorbs the endosperm.

Serology - The Study of serums and their properties.

Somatic cells - The cells of which the body of an organism is made up of as opposed to the reproductive or germ cells.

Species - A category used in classification of organisms that consists of a group of similar individuals that can breed among themselves and produce fertile offspring.

Stomata - A term used for a pore, large numbers of which are present in the epidermis of leaves, culm sheaths, and culms of bamboos. Stomata are involved in gas exchange between the plant and the atmosphere.

Sympatric - Term used for groups of similar plants, that although in close proximity and theoretically capable of interbreeding, do not interbreed, because of difference in behaviour such as flowering time.

Synapsis - Close pairing of homologs at meiosis.

Telomere - The tip or end of a chromosome.

Ultracentrifuge - A high speed centrifuge used to measure the rate of sedimentation of colloidal particles or to separate macromolecules such as proteins or nucleic acids from solutions. Utracentrifuges are electrically driven and are capable of speeds upto 60,000 rpm.

Xylarium - A collection of samples of wood specimens meant for identification and reference purpose.

INDEX